THE SENIC

COLLECTION EDITOR:

GILLDA LEITENBERG

GLOBAL MATTERS

EDITED BY IAN WALDRON AND

NANCY STEINHAUER

For my teachers,
who taught me how to search.
Nancy Steinhauer

For my students, who taught me
that the search never ends.
Ian Waldron

McGraw-Hill Ryerson Limited

Toronto • Montreal • New York • Auckland • Bogotá
Caracas • Lisbon • London • Madrid • Mexico • Milan
New Delhi • San Juan • Singapore • Sydney • Tokyo

Global Matters
The Senior Issues Collection

ISBN 0-07-551702-7

1 2 3 4 5 6 7 8 9 10   BG   4 3 2 1 0 9 8 7 6 5

Printed and bound in Canada

Canadian Cataloguing in Publication Data
Main entry under title:

Global matters

(The Senior issues collection)
ISBN 0-07-551702-7

1. Readers (Secondary). I. Waldron, Ian.
II. Steinhauer, Nancy. III. Series.

PE1121.G56 1995   808'.0427   C94–932692–5

Editor: *Kathy Evans*
Supervising Editor: *Nancy Christoffer*
Permissions Editor: *Jacqueline Donovan*
Copy Editor: *Judith Kalman*
Proofreader: *Gail Marsden*
Designer: *Mary Opper*
Typesetter: *Pages Design Ltd.*
Photo Researcher: *Elaine Freedman*
Cover Illustrator: *Marc Mongeau*

The editors wish to thank reviewer Freda Appleyard for her comments and advice.

This book was manufactured in Canada using acid-free and recycled paper.

Contents

Introduction

As we move into the twenty-first century, we are increasingly inundated with images of, and reports about, the problems our world faces: racism and other violations of human rights; war and political upheaval; economic inequities, poverty, and homelessness; hunger and other health issues; the effect of technology on our lives; and environmental destruction. In the face of such grave issues, it would be easy to lose hope and give in to despair, overwhelmed by the immensity of the task before us: the task of learning new ways to view and care for the earth and its inhabitants. Yet, time and time again in our research and preparation for this anthology, we encountered writing by and about people who were willing to grapple with these issues and wrestle them to the ground, people who have sought solutions and in some small way found them.

No anthology could adequately represent all the dangers that threaten our world. We decided that we would attempt simply to provide you with a sampling of perspectives on some of the issues that we found most compelling, help you consider new avenues of investigation, and, finally, include what evidence we found that there is reason for hope and the possibility of change.

The poems, essays, plays, and short stories we have chosen are ones that intrigued us, moved us, and made us argue and question and even change our perceptions. We have tried to include contemporary writing as well as more traditional selections. We have tried to draw from a variety of authors from a variety of origins. We have tried to represent the global community of authors, poets, and playwrights.

We have considered this process of reading, thinking, discussing, and learning a kind of journey, the beginning of an endless journey to search out and understand our lives in the age of the "global village." We hope that the result of our efforts will move you to embark upon a similar odyssey, a similar process of inquiry, thought, debate, expression, and action. We hope that your journey will move you well beyond this book into your community, and into the bigger community, the world at large.

John Guare writes in his play Six Degrees of Separation:

> ...everybody on this planet is separated by
> only six other people. Six degrees of
> separation...every person is a new door,
> opening up into other worlds. Six degrees
> of separation between me and everyone else
> on this planet.

We should realize our connection to all other inhabitants of this planet and the opportunities and obligations that this connection affords us. We hope that as you read and reflect, you will come to appreciate your place in the world, and that you will begin to understand the profound effect that your actions can, and do, have on others. Finally, we hope that you will come to realize that the personal perspective is necessarily the global perspective, and that the global is necessarily personal.

Nancy Steinhauer and Ian Waldron

Value Earth

from Value Earth Poster, 1992

BY

DONELLA H.

MEADOWS

If the world were a village of 1000 people, it would include:
- 584 Asians
- 124 Africans
- 95 East and West Europeans
- 84 Latin Americans
- 55 Soviets (including for the moment Lithuanians, Latvians, Estonians, and other national groups)
- 52 North Americans
- 6 Australians and New Zealanders

The people of the village have considerable difficulty in communicating:
- 165 people speak Mandarin
- 86 English
- 83 Hindi/Urdu
- 64 Spanish
- 58 Russian
- 37 Arabic

That list accounts for the mother tongues of only half the villagers. The other half speak (in descending order of frequency) Bengali, Portuguese, Indonesian, Japanese, German, French, and 200 other languages.

In this village of 1000 there are:
- 329 Christians (among them 187 Catholics, 84 Protestants, 31 Orthodox)
- 178 Muslims
- 167 "non-religious"
- 132 Hindus
- 60 Buddhists
- 45 atheists
- 3 Jews
- 86 all other religions

One-third (330) of the 1000 people in the world village are children and only 60 are over the age of 65. Half the children are immunized against preventable infectious diseases such as measles and polio.

Just under half of the married women in the village have access to and use modern contraceptives.

This year 28 babies will be born. Ten people will die, 3 of them for lack of food, 1 from cancer, 2 of the deaths are of babies born within the year. One person of the 1000 is infected with the HIV virus; that person most likely has not yet developed a full-blown case of AIDS.

With the 28 births and 10 deaths, the population of the village next year will be 1018.

In this 1000-person community, 200 people receive 75 percent of the income; another 200 receive only 2 percent of the income.

Only 70 people of the 1000 own an automobile (although some of the 70 own more than one automobile).

About one-third have access to clean, safe drinking water.

Of the 670 adults in the village, half are illiterate.

The village has six acres [2.4 ha] of land per person, 6000 acres [2400 ha] in all, of which
  700 acres [280 ha] are cropland
  1400 acres [560 ha] pasture
  1900 acres [760 ha] woodland
  2000 acres [800 ha] desert, tundra, pavement and other
    wasteland
The woodland is declining rapidly; the wasteland is increasing. The other land categories are roughly stable.

The village allocates 83 percent of its fertilizer to 40 percent of its cropland—that owned by the richest and best-fed 270 people. Excess fertilizer running off this land causes pollution in lakes and wells. The remaining 60 percent of the land, with its 17 percent of the fertilizer, produces 28 percent of the food grains and feeds for 73 percent of the people. The average grain yield on that land is one-third the harvest achieved by the richer villagers.

In the village of 1000 people, there are:
  5 soldiers
  7 teachers
  1 doctor
  3 refugees driven from home by war or drought

The village has a total budget each year, public and private, of over $3 million—$3000 per person if it is distributed evenly (which, we have already seen, it isn't).

Of the total $3 million:
  $181 000 goes to weapons and warfare
  $159 000 for education
  $132 000 for health care

The village has buried beneath it enough explosive power in nuclear weapons to blow itself to smithereens many times over. These weapons are under the control of just 100 of the people. The other 900 people are watching them with deep anxiety, wondering whether they can learn to get along together; and if they do, whether they might set off the weapons anyway through inattention or technical bungling; and, if they ever decide to dismantle the weapons, where in the world village they would dispose of the radioactive materials of which the weapons are made.

In It

BY

GEORGE

JOHNSTON

The world is a boat and I'm in it
Going like hell with the breeze;
Important people are in it as well
Going with me and the breeze like hell—
It's a kind of a race and we'll win it.
Out of our way, gods, please!

The world is a game and I'm in it
For the little I have, no less;
Important people are in it for more,
They watch the wheel, I watch the door.
Who was the first to begin it?
Nobody knows, but we guess.

The world is a pond and I'm in it,
In it up to my neck;
Important people are in it too,
It's deeper than this, if we only knew;
Under we go, any minute—
A swirl, some bubbles, a fleck...

Together at the Death of a Stranger

~

BY

MICHELE

McALPINE

My father died when I was a child. Like anyone who has lost a loved one, I will never forget the day as long as I live. He died in a restaurant. His car had broken down and he had gone in to buy a cup of coffee and wait for a tow truck. Precisely 10 minutes after he hung up the phone from calling my mother to say he would be late arriving home, he was dead. How could we have known at 11:10 a.m. that day that our lives had changed forever? No trumpets blared, no bands played, no balloons were released to mark the moment. He was just gone.

Fifteen years later I still recall the most intricate moments of the days that followed. The growing collection of tinfoil-clad foodstuffs from neighbours, the endless chorus of "yes, he was a fine man...," the long nights lying awake and wondering if this loss was really happening. I kept waiting for my father to come home so I could tell him about this wonderful man who had passed on so unexpectedly. But he never came home.

As the days turned into months, the healing began and now, all these years later, I remember him with a smile. If anything from the past still haunts me it is only that I wasn't there when he died. I didn't get to say goodbye. Worse than that, I didn't get to try to help him. In my mind, he died in a restaurant by himself, undoubtedly surrounded by veritable strangers, and I've never known if anyone came to his aid.

Until now. Something happened this week that gave me my answer.

As I was leaving a restaurant with a friend, an elderly man collapsed in our path. Before I realized what had happened, my friend Kevin had knelt down by the man, whispering encouragement and covering him with his coat to keep him warm. Within minutes, there was a group of passers-by who had all stopped to ask how they could help.

When I returned from calling an ambulance, the man's condition had deteriorated and life-saving measures were being taken. Even though we had our suspicions, we learned only later that, at that point, the man was dead. Not that anyone in the crowd acknowledged this, mind you, because collectively we had a mission. We were determined that this man would pull through, and a small group of us worked like a well-oiled machine to ensure it. (The fact that he was already dead was a mere inconvenience.)

We ran into the street to flag the ambulance, we made sure his eyeglasses, which had fallen from his face when he collapsed, were safely back in his pocket, we kept laying more coats over him to ensure his ultimate comfort and, most important, we kept his heart beating with our very hands. And when the ambulance whisked him off, we felt let down that we had not been allowed to climb in with him and see our challenge through. Some of us were openly emotional, while others, their heads hung low, shook hands in a show of what we had shared. We never saw the man again. He was just gone.

Kevin and I got into our car and drove away from the scene, but never really left it behind. Days later we found ourselves checking the local newspaper for an obituary we were sure would appear. It did. The man's name was Nigel Anderson; he was 87 and predeceased by his wife, Eleanor, but this "much cherished" man was survived by his daughter, Ruth. (Some of the names in this story have been changed.)

I felt compassion for Ruth. I knew she would experience all that I had gone through when my father passed away. The growing collection of tinfoil-clad foodstuffs from neighbours, the endless chorus of "yes, he was a fine man...," the long nights lying awake and wondering if the loss was really happening. I knew that for the rest of her life she would wonder just what had happened. Who had been the last person with her father, where exactly did he die and was it as awful as she imagined?

And that's why, on the day before Nigel was laid to rest, Kevin and I went to the funeral home to meet Ruth and answer her questions. Our hearts were pounding as we entered the room full of strangers. A rather frail-looking woman glanced up to acknowledge our entrance and hurried toward us. She said that even though she didn't recognize us, she knew who we were. The police had told her that people had come to the aid of her dad and that he had died outside a restaurant—was that true? And so our conversation began.

She embraced us and thanked us for our efforts. I told her that her father had died amidst a show of concern and she needn't worry that he had died alone. It occurred to me that one of the most private events of this woman's life was being shared by strangers, and while the conversation lasted only 15 minutes, I knew she would never forget it.

And likewise, I will never forget Nigel Anderson. Through him I got my own answer to the question that had haunted me for years. I know now that my father did not die alone. Someone was there to call an ambulance; to put his eyeglasses in his pocket; to keep him warm and to keep his heart beating with their hands. My father had been cared for, just as we had cared for Nigel, because apparently that is what strangers do. It was something that, in a split second, needed to be done and so we did it.

And this is astounding to me. For every day we read in the newspaper that one human being has killed another, that people are boarding New York commuter trains and murdering those in their path. We read of pestilence and disease, of families being broken apart by domestic violence. All these stories tell us exactly the same thing—that the regard for human life is seemingly diminishing before our very eyes.

All that considered, it's no wonder that the care and love shown to Nigel astounds me. Maybe it's easier to administer care to a stranger, because there is no emotional attachment and therefore you have nothing to lose. Or maybe it's selfish motives that make us extend ourselves to people in need—a way to boost our own self-worth and confirm that we are capable of reaching out to others. Whatever the case, in a world where violence seems to reign, there are glimmers of compassion and a respect for human dignity. I know, because I witnessed such a glimmer.

I'm sure Nigel Anderson—and my father—would agree.

Prayer Before Birth

BY

LOUIS MACNEICE

I am not yet born; O hear me.
Let not the bloodsucking bat or the rat or the stoat or the
    clubfooted ghoul come near me.

I am not yet born; console me.
I fear that the human race may with tall walls wall me,
    with strong drugs dope me, with wise lies lure me,
        on black racks rack me, in blood-baths roll me.

I am not yet born; provide me
With water to dandle me, grass to grow for me, trees to talk
    to me, sky to sing to me, birds and a white light
        in the back of my mind to guide me.

I am not yet born; forgive me
For the sins that in me the world shall commit, my words
    when they speak me, my thoughts when they think me,
        my treason engendered by traitors beyond me,
            my life when they murder by means of my
                hands, my death when they live me.

I am not yet born; rehearse me
In the parts I must play and the cues I must take when
    old men lecture me, bureaucrats hector me, mountains
      frown at me, lovers laugh at me, the white
        waves call me to folly and the desert calls
         me to doom and the beggar refuses
          my gift and my children curse me.

I am not yet born; O hear me.
Let not the man who is beast or who thinks he is God
    come near me.

I am not yet born; O fill me
With strength against those who would freeze my
    humanity, would dragoon me into a lethal automaton,
      would make me a cog in a machine, a thing with
        one face, a thing, and against all those
         who would dissipate my entirety, would
          blow me like thistledown hither and
          thither or hither and thither
          like water held in the
          hands would spill me

Let them not make me a stone and let them not spill me.
Otherwise kill me.

The Answer

~

BY

ROBINSON

JEFFERS

Then what is the answer?—Not to be deluded by dreams.
To know that great civilizations have broken down into
    violence, and their tyrants come, many times before.
When open violence appears, to avoid it with honor or choose
    the least ugly faction; these evils are essential.
To keep one's own integrity, be merciful and uncorrupted
    and not wish for evil; and not be duped
By dreams of universal justice or happiness. These dreams
    will not be fulfilled.
To know this, and to know that however ugly the parts appear
    the whole remains beautiful. A severed hand
Is an ugly thing, and man disseevered from the earth and the stars
    and his history...for contemplation or in fact...
Often appears atrociously ugly. Integrity is wholeness, the
    greatest beauty is
Organic wholeness, the wholeness of life and things, the
    divine beauty of the universe. Love that, not man
Apart from that, or else you will share man's pitiful confusions,
    or drown in despair when his days darken.

Nausea

BY

DI

BRANDT

we must believe it can happen we must believe
what the eye sees & the heart knows trembling
we must believe in the slow dying of the earth
the accumulation of toxic wastes plastic pop can
rings strangling sea birds poisoning fish we must
believe in the slow dying of the river where you
sat last summer watching muskrats across the
small channel digging their house of mud under
the fireweed while you searched for the precise
word to describe the shade of sky above the foot
bridge pale lemon transparent lemon orange framing
the bright purple weeds we must believe in the
fierceness of the fear that attacks us at night like
hunger like rage it is not the Earth that betrays us
just as the meaning of hunger is not in the shortage
of food & it is not our rage that will kill us but
the absence of rage walking the footpath along the
boulevard you remember the nausea you once felt
at not being a tree you remember the grove of goddesses
murky shadows dancing & holding out their hands
you understand in the glowing dusk how you are
utterly alone how you are small in the huge universe
you lift up your eyes from the gravel path with its
precise strewn pebbles to the green trees we must
believe it is possible the blood whispers urgently
to go on living we must believe in the trees

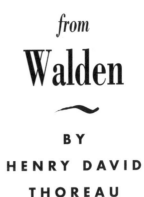

from

Walden

~

BY

HENRY DAVID

THOREAU

went to the woods because I wished to live deliberately, to front only the essential facts of life, and see if I could not learn what it had to teach, and not, when I came to die, discover that I had not lived. I did not wish to live what was not life, living is so dear; nor did I wish to practise resignation, unless it was quite necessary. I wanted to live deep and suck out all the marrow of life, to live so sturdily and Spartan-like as to put to rout all that was not life, to cut a broad swath and shave close, to drive life into a corner, and reduce it to its lowest terms, and, if it proved to be mean, why then to get the whole and genuine meanness of it, and publish its meanness to the world; or if it were sublime, to know it by experience, and be able to give a true account of it in my next excursion. For most men, it appears to me, are in a strange uncertainty about it, whether it is of the devil or of God, and have *somewhat hastily* concluded that it is the chief end of man here to "glorify God and enjoy him forever."

Still we live meanly, like ants; though the fable tells us that we were long ago changed into men; like pygmies we fight with cranes; it is error upon error, and clout upon clout, and our best virtue has for its occasion a superfluous and evitable wretchedness. Our life is frittered away by detail. An honest man has hardly need to count more than his ten fingers, or in extreme cases he may add his ten toes, and lump the rest. Simplicity, simplicity, simplicity! I say, let your affairs be as two or three, and not a hundred or a thousand; instead of a million count half a dozen, and keep your accounts on your thumb nail. In the midst of this chopping sea of civilized life, such are the clouds and storms and quicksands and thousand-and-one items to be allowed for, that a man has to live, if he would not founder and go to the bottom and not make his port at all, by dead reckoning, and he must be a great calculator indeed who succeeds. Simplify, simplify. Instead of three meals a day, if it be necessary eat but one; instead of a hundred dishes, five; and reduce other things in proportion. Our life is like a German Confederacy, made up of petty states, with its boundary forever fluctuating, so that even a German cannot tell you how it is bounded at any moment. The nation itself, with all its so called internal improvements, which, by the way, are all external and superficial, is just such an unwieldy and overgrown establishment, cluttered with furniture and tripped up by its own traps, ruined by luxury and heedless expense, by want of calculation and a worthy aim, as the million households in the land; and the only cure for it as for them is in a rigid economy, a stern and more than Spartan simplicity of life and elevation of purpose. It lives too fast. Men think that it is essential that the *Nation* have commerce, and export ice, and talk through a telegraph, and ride thirty miles an hour, without a doubt, whether *they* do or not; but whether we should live like baboons or like men, is a little uncertain. If we do not get out sleepers, and forge rails, and devote days and nights to the work, but go to tinkering upon our *lives* to improve *them*, who will build railroads? And if railroads are not built, how shall we get to heaven in season? But if we stay at home and mind our business, who will want railroads? We do not ride on the railroad; it rides upon us. Did you ever think what those sleepers are that underlie the railroad? Each one is a man, an Irish-man, or a Yankee man. The rails are laid on them, and they are covered with sand, and the cars run smoothly over them. They are sound sleepers, I assure you. And every few years a new lot is laid down and run over; so that, if some have the pleasure of riding on a rail, others have the misfortune to be ridden upon. And when they run over a man that is walking in his sleep, a supernumerary sleeper in the wrong position, and wake him up, they suddenly stop the cars, and make a hue and cry about it, as if this were an exception. I am glad to know that it takes a gang of men for every five miles to keep the sleepers down and level in their beds as it is, for this is a sign that they may sometime get up again.

Why should we live with such hurry and waste of life? We are determined to be starved before we are hungry. Men say that a stitch in time saves nine, and so they take a thousand stitches to-day to save nine to-morrow. As for *work*, we haven't any of any consequence. We have the Saint Vitus' dance, and cannot possibly keep our head still. If I should only give a few pulls at the parish bell-rope, as for a fire, that is, without setting the bell, there is hardly a man on his farm in the outskirts of Concord, notwithstanding that press of engagements which was his excuse so many times this morning, nor a boy, nor a woman, I might almost say, but would forsake all and follow that sound, not mainly to save property from the flames, but, if we will confess the truth, much more to see it burn, since burn it must, and we, be it known, did not set it on fire,—or to see it put out, and have a hand in it, if that is done as handsomely; yes, even if it were the parish church itself. Hardly a man takes a half hour's nap after dinner, but when he wakes he holds up his head and asks, "What's the news?" as if the rest of mankind had stood his sentinels. Some give directions to be waked every half hour, doubtless for no other purpose; and then, to pay for it, they tell what they have dreamed. After a night's sleep the news is as indispensable as the breakfast. "Pray tell me any thing new that has happened to a man any where on this globe",—and he reads it over his coffee and rolls, that a man has had his eyes gouged out this morning on the Wachito River; never dreaming the while that he lives in the dark unfathomed mammoth cave of this world, and has but the rudiment of an eye himself....

Let us spend one day as deliberately as Nature, and not be thrown off the track by every nutshell and mosquito's wing that falls on the rails. Let us rise early and fast, or break fast, gently and without perturbation; let company come and let company go, let the bells ring and the children cry,—determined to make a day of it. Why should we knock under and go with the stream? Let us not be upset and overwhelmed in that terrible rapid and whirlpool called a dinner, situated in the meridian shallows. Weather this danger and you are safe, for the rest of the way is down hill. With unrelaxed nerves, with morning vigor, sail by it, looking another way, tied to the mast like Ulysses. If the engine whistles, let it whistle till it is hoarse for its pains. If the bell rings, why should we run? We will consider what kind of music they are like. Let us settle ourselves, and work and wedge our feet downward through the mud and slush of opinion, and prejudice, and tradition, and delusion, and appearance, that alluvion which covers the globe, through Paris and London, through New York and Boston and Concord, through church and state, through poetry and philosophy and religion, till we come to a hard bottom and rocks in place, which we can call *reality*, and say, This is, and no mistake; and then begin, having a *point d'appui*, below freshet and frost and fire, a place where you might found a wall or a state, or set a lamp-post safely, or perhaps a gauge, not a Nilometer, but a Realometer, that future ages might know how deep a freshet of shams and appearances had gathered from time to time. If you stand right fronting and face to face to a fact, you will see the sun glimmer on both its surfaces, as if it were a cimeter, and feel its sweet edge dividing you through the heart and marrow, and so you will happily conclude your mortal career. Be it life or death, we crave only reality. If we are really dying, let us hear the rattle in our throats and feel cold in the extremities; if we are alive, let us go about our business.

Time is but the stream I go a-fishing in. I drink at it; but while I drink I see the sandy bottom and detect how shallow it is. Its thin current slides away, but eternity remains. I would drink deeper; fish in the sky, whose bottom is pebbly with stars. I cannot count one. I know not the first letter of the alphabet. I have always been regretting that I was not as wise as the day I was born. The intellect is a cleaver; it discerns and rifts its way into the secret of things. I do not wish to be any more busy with my hands than is necessary. My head is hands and feet. I feel all my best faculties concentrated in it. My instinct tells me that my head is an organ for burrowing, as some creatures use their snout and fore-paws, and with it I would mine and burrow my way through these hills. I think that the richest vein is somewhere hereabouts; so by the divining rod and thin rising vapours I judge; and here I will begin to mine.

The
Peace of
Wild Things

~

BY
WENDELL
BERRY

When despair for the world grows in me
and I wake in the night at the least sound
in fear of what my life and my children's lives may be,
I go and lie down where the wood drake
rests in his beauty on the water, and the great heron feeds.
I come into the peace of wild things
who do not tax their lives with forethought
of grief. I come into the presence of still water.
And I feel above me the day-blind stars
waiting with their light. For a time
I rest in the grace of the world, and am free.

Walking

BY

LINDA

HOGAN

t began in dark and underground weather, a slow hunger moving toward light. It grew in a dry gully beside the road where I live, a place where entire hillsides are sometimes yellow, wind-blown tides of sunflower plants. But this one was different. It was alone, and larger than the countless others who had established their lives further up the hill. This one was a traveler, a settler, and like a dream beginning in conflict, it grew where the land had been disturbed.

I saw it first in early summer. It was a green and sleeping bud, raising itself toward the sun. Ants worked around the unopened bloom, gathering aphids and sap. A few days later, it was a tender young flower, soft and new, with a pale green center and a troop of silver gray insects climbing up and down the stalk.

Over the summer this sunflower grew into a plant of incredible beauty, turning its face daily toward the sun in the most subtle of ways, the black center of it dark and alive with a deep blue light, as if flint had sparked an elemental fire there, in community with rain, mineral, mountain air, and sand.

As summer changed from green to yellow there were new visitors daily: the lace-winged insects, the bees whose legs were fat with pollen, and grasshoppers with their clattering wings and desperate hunger. There were other lives I missed, lives too small or hidden to see. It was as if this plant with its host of lives was a society, one in which moment by moment, depending on light and moisture, there was great and diverse change.

There were changes in the next larger world around the plant as well. One day I rounded a bend in the road to find the disturbing sight of a dead horse, black and still against a hillside, eyes rolled back. Another day I was nearly lifted by a wind and sandstorm so fierce and hot that I had to wait for it to pass before I could return home. On this day the faded dry petals of the sunflower were swept across the land. That was when the birds arrived to carry the new seeds to another future.

In this one plant, in one summer season, a drama of need and survival took place. Hungers were filled. Insects coupled. There was escape, exhaustion, and death. Lives touched down a moment and were gone.

I was an outsider. I only watched. I never learned the sunflower's golden language or the tongues of its citizens. I had a small understanding, nothing more than a shallow observation of the flower, insects, and birds. But they knew what to do, how to live. An old voice from somewhere, gene or cell, told the plant how to evade the pull of gravity and find its way upward, how to open. It was instinct, intuition, necessity. A certain knowing directed the seedbearing birds on paths to ancestral homelands they had never seen. They believed it. They followed.

There are other summons and calls, some even more mysterious than those commandments to birds or those survival journeys of insects. In bamboo plants, for instance, with their thin green canopy of light and golden stalks that creak in the wind. Once a century, all of a certain kind of bamboo flower on the same day. Whether they are in Malaysia or in a greenhouse in Minnesota makes no difference, nor does the age or size of the plant. They flower. Some current of an inner language passes between them, through space and separation, in ways we cannot explain in our language. They are all, somehow, one plant, each with a share of communal knowledge.

John Hay, in *The Immortal Wilderness,* has written: "There are occasions when you can hear the mysterious language of the Earth, in water, or coming through the trees, emanating from the mosses, seeping through the undercurrents of the soil, but you have to be willing to wait and receive."

Sometimes I hear it talking. The light of the sunflower was one language, but there are others, more audible. Once, in the redwood forest, I heard a beat, something like a drum or heart coming from the ground and trees and wind. That underground current stirred a kind of knowing inside me, a kinship and longing, a dream barely remembered that disappeared back to the body.

Another time, there was the booming voice of an ocean storm thundering from far out at sea, telling about what lived in the distance, about the rough water that would arrive, wave after wave revealing the disturbance at center.

Tonight I walk. I am watching the sky. I think of the people who came before me and how they knew the placement of stars in the sky, watched the moving sun long and hard enough to witness how a certain angle of light touched a stone only once a year. Without written records, they knew the gods of every night, the small, fine details of the world around them and of immensity above them.

Walking, I can almost hear the redwoods beating. And the oceans are above me here, rolling clouds, heavy and dark, considering snow. On the dry, red road, I pass the place of the sunflower, that dark and secret location where creation took place. I wonder if it will return this summer, if it will multiply and move up to the other stand of flowers in a territorial struggle.

It's winter and there is smoke from the fires. The square, lighted windows of houses are fogging over. It is a world of elemental attention, of all things working together, listening to what speaks in the blood. Whichever road I follow, I walk in the land of many gods, and they love and eat one another.

Walking, I am listening to a deeper way. Suddenly all my ancestors are behind me. Be still, they say. Watch and listen. You are the result of the love of thousands.

Exploring the Mysteries of "Deep Ecology"

~

BY

JAMES D. NATIONS

There's a movement called "deep ecology" among some environmentalists that could affect one of our planet's most precious and endangered resources: its biological diversity.

The movement is fascinating and noble—but it must be tempered with a more practical view of life on Earth. Here's why:

Plant and animal species in the world's tropical forests are being destroyed at incredible speed, in what some scientists have called an "extinction spasm." From the Amazon to Indonesia, species are being wiped out as their habitats are destroyed by farmers, loggers and others pushing into new territory. This terrible loss is eroding the genetic storehouse that scientists depend upon to develop new foods, medicines and industrial products. Chemicals and crops of the future are being lost forever.

Deep ecology deals with this problem by asserting that human beings have no right to bring other creatures to extinction or to play God by deciding which species serve us and should therefore be allowed to live. It rejects the anthropocentric view that humankind lies at the center of all that is worthwhile, saying instead that all living things—animals, plants, bacteria, viruses—have an equal inherent value.

As a conservationist, I am attracted to the core philosophy of deep ecology. Where I run into trouble with it, however, is in places like rural Central America or on the agricultural frontier in Ecuadorian Amazonia—places where human beings themselves are living on the edge of life. I have never tried to tell a Latin American farmer that he has no right to burn forest for farmland because the trees and wildlife are as inherently valuable as he and his children are. As an anthropologist and a father, I am not prepared to take on that job. You could call this the dilemma of deep ecology meeting the developing world.

When we in the industrial world stop to think about such a seemingly distant problem as this, we must resist the temptation to blame the poor farmers on the scene. The fact is that they generally understand the value of forest and wildlife better than we do in our society of microwave ovens and plastic money. They, not us, gather edible fruit, wild animals for protein, fiber for clothing and ropes, incense for religious ceremonies, natural insecticides, wood for houses, and medicinal plants.

I stayed once in southeastern Mexico with a Mayan farmer who expressed his view this way: "The outsiders come into our forest and they cut the mahogany and kill the birds and burn everything. I think they hate the forest. But I plant my crops and weed them, and I watch the animals.... As for me, I guard the forest."

Today, that Mayan farmer lives in a small remnant of rainforest surrounded by the fields and cattle pastures of 100 000 immigrant colonists. The colonists are fine people who are quick to invite you to share their meager meal. But any of us in the industrial world who want to talk with them about protecting the biological diversity that still surrounds them had better be prepared to explain how it will affect them directly.

If you tell a frontier farmer that he must not clear forest or hunt in a wildlife reserve because this threatens the planet's biological diversity, he will politely perform the cultural equivalent of rolling his eyes and saying "Sure." The same can be said for the government planner in the nation where the pioneer farmer lives.

In other words, deep ecology makes interesting conversation over the seminar table, but it won't fly on the agricultural frontier of the Third World or in the boardrooms of international development banks.

Fortunately, even as advocates speak of deep ecology, many scientists and conservationists are showing a growing pragmatism about the situation. They are recognizing that to obtain the long-term benefits of biological diversity one must focus first—or at least simultaneously—on the immediate, short-term needs of individual people. Few wild gene pools are likely to survive intact in places where people have to struggle simply to provide their basic, daily needs.

This recognition must guide efforts to save the world's tropical forests. People on the frontiers of the developing world must receive material incentives that allow them to prosper by protecting biological diversity rather than by destroying it, and their welfare must be given equal consideration with the welfare of the forests. That done, we can return to the aesthetic arguments of deep ecology with the knowledge that, when we look up from our discussion, there will still be biological diversity left to experience and enjoy.

those subjected to radioactivity who did not die were marked for life

~

BY

KRISTJANA
GUNNARS

the idea of storing
nuclear waste in the sea once had
a strong following
but i understand
there is not enough water in all the oceans
to cover even ten percent of our sins

today a black-capped bird
eats flies at the landing place
a young gull circles
with a rasping call

the lumpsucker season started
the day before yesterday
freshly caught fish lie
in a wagon outside the shed
and a little girl sells them, her hands folded

i hear
the banks from skagatá to langanes are open
and northern boats are out on the cold sea
too cold, they say
and the lumpsucker is late this year

again sounds the za-za of the gull
down by the broken boards
and rusty nails
today i feel particularly aware
of the possibility

that radioactivity resembles sin
it spreads unseen
through plankton, algae, fish
and i have come to buy it from a little girl
with folded hands

The Clean Answer to King Coal's Poisonous Reign

BY

ANDREW KENNY

I am a vegetarian and therefore a supporter of nuclear power. I became a vegetarian 20 years ago because I believed that the farming and slaughter of animals was cruel and because I thought it was a shameful waste of food to take the grain that could feed ten people and give it to a beast to produce only enough meat to feed one person. My reasons, in short, for turning to vegetarianism were respect for humans and nature. These are precisely my reasons for turning to nuclear power.

Rather late in life I changed careers, became an engineer and joined an electricity utility. They required me to spend one and a half years at a coal station before I could join the nuclear section. My stay at the coal station shocked me and greatly increased my zeal for nuclear power. I have taken a vow to myself that if I ever have children I shall not allow them to grow up near a coal station (I should be perfectly happy for them to grow up near a nuclear one). But distance from coal power is not enough to escape its menace.

I support nuclear power for one reason only: that it is cleaner and safer than other practicable large-scale sources of electricity. The best way to see this is in the question of waste. In the case of nuclear power, a small amount of uranium is dug out of the ground, refined, passed through a nuclear reactor, stored and then returned to the ground again. The original uranium lying in the ground is mildly dangerous in that it emits a radioactive gas, radon, that naturally seeps out of the ground, sometimes causing lung cancer. The nuclear waste is dangerous in the short term but in the long term, because of its shorter half-lives, it is less dangerous than what it came from.

In the case of coal power, a huge amount of coal is dug out of the ground, passed through a coal furnace and converted into dangerous substances which are then either poured into the atmosphere for plants and humans to breathe or dumped on-to ash tips, leaching their poisons into the water courses. The original coal is quite safe. The coal wastes are very dangerous, and unlike the nuclear wastes, many remain dangerous for ever.

Nuclear power produces only one form of pollution, radiation, and only two possible dangers, cancer and genetic disease. Even in these it is overshadowed by coal. Large amounts of radiation can certainly cause cancer but low levels of radiation are a nat-ural and inescapable fact of life. Soil, milk, stone, wood, flesh—these are all radioactive and with the sun they give us a "back-ground radiation" massively larger than any radiation received even on the doorstep of a nuclear station.

Coal, too, is radioactive and, having heard that a coal station routinely emits more radiation than a nuclear one, I put radiation badges on the workers of our coal station and measured the levels there. Sure enough, the radiation next to the coal station was twice as high as that next to our nuclear station, although the levels were both very low.

Far more serious are the chemical cancer agents from coal. In the short term I lived in the small township at the coal station, two people, one the power station manager's wife, died grim deaths by cancer. Their deaths are doubtless of no statistical significance but the carcinogens in the coal wastes are a hard fact. They include organic carcinogens such as nitrosamines and benzopyrenes and, worse, heavy metals such as cadmium and arsenic.

Cadmium is normally locked safely into the coal but when the coal is burnt in the furnace the cadmium vaporizes, turns into tiny particles, passes through the smokestacks and spreads through the atmosphere to settle finely on the ground where it dis-solves in water, enters human tissue and causes cancer. Cadmium has a half-life of infini-ty. When Chernobyl's ruined reactor has become less radioac-tive than the soil in your back garden, when the Pharoahs' mighty pyramids have crumbled into sand, when our sun has become a Red Giant and boiled our oceans dry, the cadmium from your local coal station will be as deadly as on the day it left the smokestack.

Radiation in large amounts is known to cause genetic damage in animals, but the curious fact is that it has never been observed to cause genetic damage in human beings. The first generation of survivors after the bombs on Hiroshima and Nagasaki was minutely studied but the babies born to them showed not the slightest increase in abnormality. Coal power stations emit many chemicals, such as the polycyclic aromatic hydrocarbons (PAH), known to cause genetic defects in animals. They may do the same in humans but this is not yet known. Those cartoons of two-headed children, so liked by the antinuclear brigade, apply at least as well to coal power.

By every measure the death rate from coal power is much larger than the death rate from nuclear power. Per unit of energy extracted, coal mining claims more than ten times as many deaths as uranium mining. Far worse are the civilian casualties. Study after study into the deaths caused by coal pollution agree that the figure is about 50 deaths per medium-sized coal station per year. This figure is necessarily tentative because coal pollution is diffuse and insidious in its effects. It translates into 25 000 deaths per year in America and 1700 deaths per year in Britain.

The World Health Organization estimates that the Chernobyl accident may cause 1600 premature deaths over the next 30 years. Thus coal power in the United Kingdom kills more civilians in one year than Chernobyl will kill in 30. And Chernobyl is the only civilian nuclear accident ever to claim a life. Three Mile Island killed nobody, injured nobody and exposed the nearest civilians to radiation less than one tenth of one dental X-ray.

The direct human casualties from coal power, although far larger than from nuclear power, are dwarfed by the devastation coal causes to the environment. Acid rain, caused by sulphur oxides (SOx) and nitrogen oxides (NOx) from coal stations and other burning of fossil fuel, has already caused vast damage to the planet's lakes and forests and the damage is spreading.

Even more ominous are the future consequences of the millions of tons of carbon dioxide that coal stations pour into the atmosphere. The "greenhouse effect" has been much in the news recently. The global weather system is very complicated and very finely poised, and it is difficult to make short-term assessments.

What is absolutely certain is that the fragile balance depends crucially on the amount of carbon dioxide in the atmosphere and that amount is rising inexorably. We are sliding towards some immense change, perhaps a catastrophe. The human race is threatened by several rising trends—population growth, the demand for resources—but of all the graphs of doom none gives such apocalyptic warning as the rising level of carbon dioxide. The future of our civilization may well depend on reversing it. Nuclear stations produce not one drop of acid rain, not one breath of carbon dioxide.

There is no such thing as a clean coal power station. A "clean stack" simply means that the visible pollution has been removed. But most coal pollution is invisible. Only a minority of advanced coal stations have chemical scrubbing and even those only attempt to reduce the SOx and NOx. None try to to remove the heavy metals. It is impossible to remove the carbon dioxide.

The most fanatical antagonists of nuclear power are forced to admit that its safety record is without equal in power generation. But, they cry, what if the really big nuclear power disaster happens? The really big disaster will not happen because it cannot happen. No nuclear power accident can match the damage done by the routine operation of coal power stations.

The superb safety record of nuclear power in the West is not because of any superhuman diligence by nuclear engineers—indeed many of them were hair-raisingly sloppy in the early days—but because the designs they have chosen are intrinsically safe. Chernobyl happened because the RMBK reactor was not intrinsically safe.

Nuclear power, unlike any other large industrial process I can think of, offers itself to inherent safety. There are nuclear reactor designs now on the drawing board in which safety is entirely passive: human operators will control the reactor only while it is running within safe limits; if it deviates from the limits, the laws of nature will overrule the operators and shut the reactor down safely. Nuclear power, safe now, offers yet more safety in the future.

A paradoxical reason for the fear of nuclear power is its unique ability to take precautions. Precautions scare people.

An ambulance parked conspicuously on the beach and marked "shark attack unit" would frighten bathers rather than reassure them. Nuclear power is able to take precautions and does so; coal is unable to take them and does not.

When you have been working near the reactor, you are scanned for radioactive contamination because it is so easy to do so; when you have been working in coal, you are not examined for particle penetration of the deep lung tissue because it is difficult to do so. It is quite easy to collect and store nuclear waste and so it is done. It is impossible to collect and store coal waste and so it is not done. The nuclear safety measures are highly visible and make people nervous.

There are passionate and articulate pro-nuclear voices within the nuclear industry but these voices are deliberately silenced by the industry itself. This was my biggest surprise on entering it. But it is easy to see why. The best argument for nuclear power is simply a comparison between its dangers and the dangers of all competitors, including the renewable sources, but mainly coal.

Nuclear stations usually belong to utilities, such as the CEGB, which make most of their electricity from coal. The dangers of coal are much greater than those of nuclear power but less well publicized and of course the utilities want to keep it that way. When I first joined our nuclear section, I was asked to write a publicity blurb to attract young people to a career in nuclear power. I did so, and included a comparison between nuclear waste and coal waste. My blurb was published in full, except that all reference to coal waste was removed.

Some people suspect that the power utilities are covering up a big secret about nuclear power. They are right. The big secret is that nuclear power is very much safer than coal.

Indeed if I were an engineer in the CEGB, I would be in trouble for this article. This is why I am coyly avoiding naming my country or utility. If privatization in Britain were to mean the nuclear power stations becoming independent of the coal ones and competing against them, then the proponents of nuclear power would be free to go on to the attack and the British people would soon hear what the people in the industry already know, that coal stations are ecological time bombs.

I am sure that coal's magnificent machines will remain in the folk memory long after they have been condemned to scrapyards and museums.

Nothing in nuclear power can match the romance and splendour of the coal age. Nuclear power is a bit of a bore but it has arrived, perhaps at the eleventh hour, to offer us salvation from looming ecological disaster. If we do not take up its offer, our grandchildren and their grandchildren will neither understand us nor forgive us.

Lessons From the Third World:

Sometimes the simplest ideas turn out to be the best ones

BY

BEN

BARBER

There were 15 minutes left before the train from Bombay to Delhi was to leave, so I decided to buy a cup of "cha," sweet, milky Indian tea. I gulped the brew served in a brown clay cup in order to return the container before catching the train. Looking around, however, I noticed other customers simply dashing their empty cups to the ground where they were quickly crushed into the red-brown earth by the throngs. These were disposable cups—Indian style.

A month later, I bought coffee to go at Washington, DC's, Union Station. It came in a plastic foam cup that will take hundreds of years to decompose in a landfill, all the while giving off ozone-destroying chemicals. After I drank the coffee, I tossed the cup into a plastic trash bag to join a growing collection of serving items destined for the dump.

I wondered, how could we make the technology flow—typically from West to East—a more even exchange, with the planet as beneficiary? Would my children one day buy drinks, sweets, or even compact discs in packages made from leaves, bamboo, or other materials that decay into a rich compost? "There's a big question as to what is adaptable from the Third World," remarks researcher John Young of the Worldwatch Institute. "Much of it takes low wages and a need for materials to justify. Sometimes you can extract a kernel of Third World wisdom."

Already some McDonald's franchises have cut down on foam packaging. Paper cups are making a comeback, and many cities are introducing recycling for newspapers, plastic, aluminum, and glass. Though these may not look like it, Young says, they are Third World solutions to ecological problems.

While most of us pay high prices for tasteless vegetables trucked from one coast to the other, the Chinese city of Shanghai produces all of its own vegetables, fertilizing them with human waste, and exports a surplus. In a similar vein, Washington, DC, has begun packaging sewage sludge for lawn fertilizer.

Still, our ingrained garbage habits will die hard. "It's a mind-set, a mentality," says Steve Hirsch of Volunteers in Technical Assistance. "Every time I go to a garage sale, I see people throwing away things that are considered gold in developing countries." Recently in Marrakesh, Morocco, Hirsch says he lost the key to his bicycle lock and visited a locksmith to have the lock cut away from the chain. Instead, the locksmith drilled four small holes, removed the lock's innards, made a new key, and reassembled the lock in about ten minutes. "Where could you have done that in America?" asks Hirsch.

In fact, technology these days flows almost exclusively from the West to developing countries: satellites, miracle rice, bypass surgery, and plastic bags. Yet India, Africa, China, and Latin America remain storehouses of knowledge that might help save the planet from the glut of its own garbage.

Each time I throw a plastic plate in the trash, I remember how I tossed the banana leaf from which I had eaten my rice to the ground in Madras, India, and saw it immediately chewed up by a gentle cow. And I remember the Artibonite Valley of Haiti where not a speck of plastic, metal, or glass spoils the earthen paths, fields, and court-yards.

So far I am aware of no systematic search for appropriate, cost-effective, Third World ideas that can be adapted in the West to preserve the planet. But as the barges loaded with garbage find fewer places to dump their loads, some of the oldest ideas on earth may turn out to be the most important.

It Is Important

~

BY

GAIL

TREMBLAY

On dark nights, when thoughts fly like nightbirds
looking for prey, it is important to remember
to bless with names every creature that comes
to mind; to sing a thankful song and hold
the magic of the whole creation close in the heart,
to watch light dance and know the sacred is alive.
On dark nights, when owls watch, their eyes
gleaming in the black expanse of starless sky,
it is important to gather the medicine bones,
the eagle feathers, the tobacco bundles, the braided
sweetgrass, the cedar, and the sage, and pray
the world will heal and breath feed the plants
that care for the nations keeping the circle whole.

On dark nights, when those who think only of themselves
conjure over stones and sing spells to feed their wills
it is important to give gifts and to love everything
that shows itself as good. It is time to turn
to the Great Mystery and know the Grandfathers have
mercy on us that we may help the people to survive.
On dark nights, when confusion makes those who envy
hate and curse the winds, face the four directions
and mumble names, it is important to stand
and see that our only work is to give what others
need, that everything that touches us is a holy
gift to teach us we are loved. When sun rises
and light surrounds life making blessings grow,
it is important to praise its coming, and exhale
letting all we hold inside our lungs travel east
and mix its power with the air; it is important to praise
dawn's power breathing in and know we live in good
relation to all creation and sing what must be sung.

"Regular Guy" Becomes a Champion

~

BY

FRANK JONES

Today, I'd like you to meet David Grassby, a 16-year-old Thornhill, Ontario, boy who never won any top academic prizes and who once played baseball for a team that was famous for being dead last. Even his dad Gerry describes David as "an ordinary boy."

He was ordinary, that is, until his teacher at St. Anthony Separate School, which he was attending in 1989 when he was 12, handed out an assignment to come up with a science fair project.

Since then, David has appeared on radio and television numerous times, his name opens the doors of mayors and oil company presidents, and he has even had a play written about him, *The Champion of the Oakbank Pond.*

Drive a little west of old Thornhill and you'll see the pond. A historic plaque reminds us that J.E.H. MacDonald, one of the Group of Seven, lived on its banks from 1913 to 1932. With willows dipping their fronds in the water and ducks paddling by with their families, it seems a scene devised for a painter. Except for the monster homes which have been allowed to intrude, in one case, right to the edge of the water. And they are only the visible part of the problem.

David learned to skate on Oakbank Pond. When David heard the pond's very existence was threatened, it was the cue for his science project. He studied first how ponds maintain themselves, then dived into hefty reports that had already been prepared for the town of Vaughan—the kind of reports that flood across politicians' desks every week on their way to oblivion.

David boiled it down to four problems:

- Salt from nearby roads in winter was killing the zooplankton, the minute creatures which help to keep the water clear.

- Lawn fertilizers from nearby gardens were leaching into the ground water and causing an unsightly algae buildup.
- Because people were feeding the ducks, more and more ducks were arriving, their waste contaminating the water.
- The foundations for the big new houses, plus storm sewers, were cutting off the ground water that kept the pond from becoming stagnant.

David talked to environmentalist Pearl Shore, who was also concerned about the pond's future, and came up with answers to each of the problems:

- Sand should be used instead of salt on roads near the pond.
- People should switch to slow-release fertilizers on their lawns.
- Signs should be put up urging people not to feed the ducks.
- The ground water question was the hardest. David's suggestion: Vaughan council should force developers to put in storm sewers in ways which would not choke off the pond.

At the town's annual winter festival at the pond, David handed out questionnaires detailing the problems and asking people what they were prepared to do. Nearly everyone agreed to tell their friends about the pond, to write to the mayor, and to personally stop feeding the ducks.

David himself wrote to Mayor Lorna Jackson. His name was soon to become familiar to her. Soon he was calling her up, asking when the signs about the ducks were going up. Today you'll see them there, a tribute to David's persistence. The town also agreed to spend $700 000 over four years to save the pond. So far, wells have been drilled along the shoreline.

Why did the politicians jump? They'd probably argue they were going to save the pond anyway. The fact is that Lesley Simpson, then a reporter for *The Star,* spotted David handing out his questionnaires at the winter carnival and wrote about him. Soon David was making regular appearances on CBC radio programs. One morning he got up at 5 a.m. with his dad and went to the CITY-TV studio to tell his story on the morning show.

The biggest thrill came when one of his heroes, environmentalist David Suzuki, came to Thornhill to film a sequence with David at Oakbank Pond for a TV special. A signed photograph of himself with Suzuki is one of his proudest possessions.

No one was more amazed at all this than his parents, Donna and Gerry. David was never the guy who spoke up in class or stood out in any special way.

"He was just like his father—a real, regular guy," says Gerry. "We were amazed that he had this spark, that he was not going to be put off by people."

Playwright Jim Betts heard David on the radio one day and thought he'd be an ideal subject for a musical for Theatre on the Move. Based at Black Creek Pioneer Village, the troupe tours schools. A lengthy collaboration with David followed, culminating in a highly successful tour of 110 schools in Spring, 1993.

Donna shed a tear the first time she saw it.

"It suddenly occurred to me what he had actually done," she says. "How many people have a song or a play written about them!"

You'd expect a boy getting that kind of exposure to become, well, just a bit full of himself. Not David.

"I always tried to stick to the original purpose: getting the community to save the pond," he says.

Another break came his way when his principal at St. Robert Catholic high school, Gerry Brand, permitted him to take a week off to tour with the play. He was asked to sign autographs and one eighth grader told him "We're really proud of you."

But David was getting the most excitement backstage. Theatre lighting is his hobby, and he took me down the basement to show me a lighting system he invented and wired himself. So when they told him he could help the stage manager, it almost beat the pleasure of seeing actor Drew Carnwath playing him onstage. Almost.

The musical launches with a song called "Anyone can make a difference." Everything that's happened, says Gerry, a business consultant, "has been a gift to David in terms of his self-confidence and his recognition that he is a person who counts, who has something to give to his community and his friends. You can be an ordinary guy and still make a contribution."

By the Waters of Babylon

BY

STEPHEN VINCENT
BENET

~

he north and the west and the south are good hunting ground,
but it is forbidden to go east. It is forbidden to go to any of the
Dead Places except to search for metal and then he who touches
the metal must be a priest or the son of a priest. Afterwards, both the
man and the metal must be purified. These are the rules and the laws;
they are well made. It is forbidden to cross the great river and look
upon the place that was the Place of the Gods—this is most strictly forbidden. We do not even say its name though we know its name. It is
there that spirits live, and demons—it is there that there are the ashes
of the Great Burning. These things are forbidden—they have been forbidden since the beginning of time.

My father is a priest; I am the son of a priest. I have been in the
Dead Places near us, with my father—at first, I was afraid. When my
father went into the house to search for the metal, I stood by the door
and my heart felt small and weak. It was a dead man's house, a spirit
house. It did not have the smell of man, though there were old bones
in a corner. But it is not fitting that a priest's son should show fear. I
looked at the bones in the shadow and kept my voice still.

Then my father came out with the metal—a good, strong piece. He looked at me with both eyes but I had not run away. He gave me the metal to hold—I took it and did not die. So he knew that I was truly his son and would be a priest in my time. That was when I was very young—nevertheless, my brothers would not have done it, though they are good hunters. After that, they gave me the good piece of meat and the warm corner by the fire. My father watched over me—he was glad that I should be a priest. But when I boasted or wept without a reason, he punished me more strictly than my brothers. That was right.

After a time, I myself was allowed to go into the dead houses and search for metal. So I learned the ways of those houses—and if I saw bones, I was no longer afraid. The bones are light and old—sometimes they will fall into dust if you touch them. But that is a great sin.

I was taught the chants and spells—I was taught how to stop the running of blood from a wound and many secrets. A priest must know many secrets—that was what my father said. If the hunters think we do all things by chants and spells, they may believe so—it does not hurt them. I was taught how to read in the old books and how to make the old writings—that was hard and took a long time. My knowledge made me happy—it was like a fire in my heart. Most of all, I liked to hear of the Old Days and the stories of the gods. I asked myself many questions that I could not answer, but it was good to ask them. At night, I would lie awake and listen to the wind—it seemed to me that it was the voice of the gods as they flew through the air.

We are not ignorant like the Forest People—our women spin wool on the wheel, our priests wear a white robe. We do not eat grubs from the tree, we have not forgotten the old writings, although they are hard to understand. Nevertheless, my knowledge and my lack of knowledge burned in me—I wished to know more. When I was a man at last, I came to my father and said, "It is time for me to go on my journey. Give me your leave."

He looked at me for a long time, stroking his beard, then he said at last, "Yes. It is time." That night, in the house of the priesthood, I asked for and received purification. My body hurt but my spirit was a cool stone. It was my father himself who questioned me about my dreams.

He bade me look into the smoke of the fire and see—I saw and told what I saw. It was what I have always seen—a river, and, beyond it, a great Dead Place and in it the gods walking. I have always thought about that. His eyes were stern when I told him—he was no longer my father but a priest. He said, "This is a strong dream."

"It is mine," I said, while the smoke waved and my head felt light. They were singing the Star song in the outer chamber and it was like the buzzing of bees in my head.

He asked me how the gods were dressed and I told him how they were dressed. We know how they were dressed from the book, but I saw them as if they were before me. When I had finished, he threw the sticks three times and studied them as they fell.

"This is a very strong dream," he said. "It may eat you up."

"I am not afraid," I said and looked at him with both eyes. My voice sounded thin in my ears but that was because of the smoke.

He touched me on the breast and the forehead. He gave me the bow and the three arrows.

"Take them," he said. "It is forbidden to travel east. It is forbidden to cross the river. It is forbidden to go to the Place of the Gods. All these things are forbidden."

"All these things are forbidden," I said, but it was my voice that spoke and not my spirit. He looked at me again.

"My son," he said. "Once I had young dreams. If your dreams do not eat you up, you may be a great priest. If they eat you, you are still my son. Now go on your journey."

I went fasting, as is the law. My body hurt but not my heart. When the dawn came, I was out of sight of the village. I prayed and purified myself, waiting for a sign. The sign was an eagle. It flew east.

Sometimes signs are sent by bad spirits. I waited again on the flat rock, fasting, taking no food. I was very still—I could feel the sky above me and the earth beneath. I waited till the sun was beginning to sink. Then three deer passed in the valley, going east—they did not wind me or see me. There was a white fawn with them—a very great sign.

I followed them, at a distance, waiting for what would happen. My heart was troubled about going east, yet I knew that I must go. My head hummed with my fasting—I did not even see the panther spring upon the white fawn. But, before I knew it, the bow was in my hand. I shouted and the panther lifted his head from the fawn. It is not easy to kill a panther with one arrow but the arrow went through his eye and into his brain. He died as he tried to spring—he rolled over, tearing at the ground. Then I knew I was meant to go east—I knew that was my journey. When the night came, I made my fire and roasted meat.

It is eight suns' journey to the east and a man passes by many Dead Places. The Forest People are afraid of them but I am not. Once I made my fire on the edge of a Dead Place at night and, next morning, in the dead house, I found a good knife, little rusted. That was small to what came afterward but it made my heart feel big. Always when I looked for game, it was in front of my arrow, and twice I passed hunting parties of the Forest People without their knowing. So I knew my magic was strong and my journey clean, in spite of the law.

Toward the setting of the eighth sun, I came to the banks of the great river. It was half-a-day's journey after I had left the god-road— we do not use the god-roads now for they are falling apart into great blocks of stone, and the forest is safer going. A long way off, I had seen water through trees but the trees were thick. At last, I came out upon an open place at the top of a cliff. There was a great river below, like a giant in the sun. It is very long, very wide. It could eat all the streams we know and still be thirsty. Its name is Ou-dis-sun, the Sacred, the Long. No man of my tribe had seen it, not even my father, the priest. It was magic and I prayed.

Then I raised my eyes and looked south. It was there, the Place of the Gods.

How can I tell what it was like—you do not know. It was there, in the red light, and they were too big to be houses. It was there with the red light upon it, mighty and ruined. I knew that in another moment the gods would see me. I covered my eyes with my hands and crept back into the forest.

Surely, that was enough to do, and live. Surely it was enough to spend the night upon the cliff. The Forest People themselves do not come near. Yet, all through the night, I knew that I should have to cross the river and walk in the Place of the Gods, although the gods ate me up. My magic did not help me at all and yet there was a fire in my bowels, a fire in my mind. When the sun rose, I thought, "My journey has been clean. Now I will go home from my journey." But, even as I thought so, I knew I could not. If I went to the Place of the Gods, I would surely die, but, if I did not go, I could never be at peace with my spirit again. It is better to lose one's life than one's spirit, if one is a priest and the son of a priest.

Nevertheless, as I made the raft, the tears ran out of my eyes. The Forest People could have killed me without fight, if they had come upon me then, but they did not come. When the raft was made, I said the sayings for the dead and painted myself for death. My heart was cold as a frog and my knees like water, but the burning in my mind would not let me have peace. As I pushed the raft from the shore, I began my death song—I had the right. It was a fine song.

"I am John, son of John," I sang. "My people are the Hill People. They are the men.
I go into the Dead Places but I am not slain.
I take the metal from the Dead Places but I am not blasted.
I travel upon the god-roads and am not afraid. E-yah! I have killed the panther, I have killed the fawn!
E-yah! I have come to the great river. No man has come there before.
It is forbidden to go east, but I have gone, forbidden to go on the great river, but I am there.
Open your hearts, your spirit, and hear my song.
Now I go to the Place of the Gods, I shall not return.
My body is painted for death and my limbs weak, but my heart is big as I go to the Place of the Gods!"

All the same, when I came to the Place of the Gods, I was afraid, afraid. The current of the great river is very strong—it gripped my raft with its hands. That was magic, for the river itself is wide and calm. I could feel evil spirits about me, in the bright morning; I could feel their breath on my neck as I was swept down the stream. Never have I been so much alone—I tried to think of my knowledge, but it was a squirrel's heap of winter nuts. There was no strength in my knowledge any more and I felt small and naked as a new-hatched bird—alone upon the great river, the servant of the gods.

Yet, after a while, my eyes were opened and I saw. I saw both banks of the river—I saw that once there had been god-roads across it, though now they were broken and fallen like broken vines. Very great they were and wonderful and broken—broken in the time of the Great Burning when the fire fell out of the sky. And always the current took me nearer to the Place of the Gods, and the huge ruins rose before my eyes.

I do not know the customs of rivers—we are the People of the Hills. I tried to guide my raft with the pole but it spun around. I thought the river meant to take me past the Place of the Gods and out into the Bitter Water of the legends. I grew angry then—my heart felt strong. I said aloud, "I am a priest and the son of a priest!" The gods heard me—they showed me how to paddle with the pole on one side of the raft. The current changed itself—I drew near to the Place of the Gods.

When I was very near, my raft struck and turned over. I can swim in our lakes—I swam to the shore. There was a great spike of rusted metal sticking out into the river—I hauled myself up upon it and sat there, panting. I had saved my bow and two arrows and the knife I found in the Dead Place but that was all. My raft went whirling downstream toward the Bitter Water. I looked after it, and thought if it had trod me under, at least I would be safely dead. Nevertheless, when I had dried my bow-string and re-strung it, I walked forward to the Place of the Gods.

It felt like ground underfoot; it did not burn me. It is not true what some of the tales say, that the ground there burns forever, for I have been there. Here and there were the marks and stains of the Great Burning, on the ruins, that is true. But they were old marks and old stains. It is not true either, what some of our priests say, that it is an island covered with fogs and enchantments. It is not. It is a great Dead Place—greater than any Dead Place we know. Everywhere in it there are god-roads, though most are cracked and broken. Everywhere there are the ruins of the high towers of the gods.

How shall I tell what I saw? I went carefully, my strung bow in my hand, my skin ready for danger. There should have been the wailings of spirits and the shrieks of demons, but there were not. It was very silent and sunny where I had landed—the wind and the rain and the birds that drop seeds had done their work—the grass grew in the cracks of the broken stone. It is a fair island—no wonder the gods built there. If I had come there, a god, I also would have built.

How shall I tell what I saw? The towers are not all broken—here and there one still stands, like a great tree in a forest, and the birds nest high. But the towers themselves look blind, for the gods are gone. I saw a fish-hawk, catching fish in the river. I saw a little dance of white butterflies over a great heap of broken stones and columns. I went there and looked about me—there was a carved stone with cut-letters, broken in half. I can read letters but I could not understand these. They said UBTREAS. There was also the shattered image of a man or a god. It had been made of white stone and he wore his hair tied back like a woman's. His name was ASHING, as I read on the cracked half of a stone. I thought it wise to pray to ASHING, though I do not know that god.

How shall I tell what I saw? There was no smell of man left, on stone or metal. Nor were there many trees in that wilderness of stone. There are many pigeons, nesting and dropping in the towers—the gods must have loved them, or, perhaps, they used them for sacrifices. There are wild cats that roam the god-roads, green-eyed, unafraid of man. At night, they wail like demons but they are not demons. The wild dogs are more dangerous, for they hunt in a pack, but them I did not meet till later. Everywhere there are the carved stones, carved with magical numbers or words.

I went North—I did not try to hide myself. When a god or a demon saw me, then I would die, but meanwhile I was no longer afraid. My hunger for knowledge burned in me—there was so much that I could not understand. After awhile, I knew that my belly was hungry. I could have hunted for my meat, but I did not hunt. It is known that the gods did not hunt as we do—they got their food from enchanted boxes and jars. Sometimes these are still found in the Dead Places—once, when I was a child and foolish, I opened such a jar and tasted it and found the food sweet. But my father found out and punished me for it strictly, for, often, that food is death. Now, though, I had long gone past what was forbidden, and I entered the likeliest towers, looking for the food of the gods.

I found it at last in the ruins of a great temple in the mid-city. A mighty temple it must have been, for the roof was painted like the sky at night with its stars—that much I could see, though the colours were faint and dim. It went down into great caves and tunnels—perhaps they kept their slaves there. But when I started to climb down, I heard the squeaking of rats, so I did not go—rats are unclean, and there must have been many tribes of them, from the squeaking. But near there, I found food, in the heart of a ruin, behind a door that still opened. I ate only the fruits from the jars—they had a very sweet taste. There was drink, too, in bottles of glass—the drink of the gods was strong and made my head swim. After I had eaten and drunk, I slept on the top of a stone, my bow at my side.

When I awoke, the sun was low. Looking down from where I lay, I saw a dog sitting on his haunches. His tongue was hanging out of his mouth; he looked as if he were laughing. He was a big dog, with a gray-brown coat, as big as a wolf. I sprang up and shouted at him but he did not move—he just sat there as if he were laughing. I did not like that. When I reached for a stone to throw, he moved swiftly out of the way of the stone. He was not afraid of me; he looked at me as if I were meat. No doubt I could have killed him with an arrow, but I did not know if there were others. Moreover, night was falling.

I looked about me—not far away there was a great, broken god-road, leading North. The towers were high enough, but not so high, and while many of the dead-houses were wrecked, there were some that stood. I went toward this god-road, keeping to the heights of the ruins, while the dog followed. When I had reached the god-road, I saw that there were others behind him. If I had slept later, they would have come upon me asleep and torn out my throat. As it was, they were sure enough of me; they did not hurry. When I went into the dead-house, they kept watch at the entrance—doubtless they thought they would have a fine hunt. But a dog cannot open a door and I knew, from the books, that the gods did not like to live on the ground but on high.

I had just found a door I could open when the dogs decided to rush. Ha! They were surprised when I shut the door in their faces—it was a good door, of strong metal. I could hear their foolish baying beyond it but I did not stop to answer them. I was in darkness—I found stairs and climbed. There were many stairs, turning around till my head was dizzy. At the top was another door—I found the knob and opened it. I was in a long small chamber—on one side of it was a bronze door that could not be opened, for it had no handle. Perhaps there was a magic word to open it but I did not have the word. I turned to the door in the opposite side of the wall. The lock of it was broken and I opened it and went in.

Within, there was a place of great riches. The god who lived there must have been a powerful god. The first room was a small ante-room—I waited there for some time, telling the spirits of the place that I came in peace and not as a robber. When it seemed to me that they had had time to hear me, I went on. Ah, what riches! Few, even, of the windows had been broken—it was all as it had been. The great windows that looked over the city had not been broken at all though they were dusty and streaked with many years. There were coverings on the floors, the colours not greatly faded, and the chairs were soft and deep. There were pictures upon the walls, very strange, very wonder-ful—I remember one of a bunch of flowers in a jar—if you came close to it, you could see nothing but bits of colour, but if you stood away from it, the flowers might have been picked yesterday. It made my heart feel strange to look at this picture—and to look at the figure of a bird, in some hard clay, on a table and see it so like our birds.

Everywhere there were books and writings, many in tongues that I could not read. The god who lived there must have been a wise god and full of knowledge. I felt I had right there, as I sought knowledge also.

Nevertheless, it was strange. There was a washing-place but no water—perhaps the gods washed in air. There was a cooking-place but no wood and though there was a machine to cook food, there was no place to put fire in it. Nor were there candles or lamps—there were things that looked like lamps but they had neither oil nor wick. All these things were magic, but I touched them and lived—the magic had gone out of them. Let me tell one thing to show. In the washing-place, a thing said "Hot" but it was not hot to the touch—another thing said "Cold" but it was not cold. This must have been a strong magic but the magic was gone. I do not understand—they had ways—I wish that I knew.

It was close and dry and dusty in their house of the gods. I have said the magic was gone but that is not true—it had gone from the magic things but it had not gone from the place. I felt the spirits about me, weighing upon me. Nor had I ever slept in a Dead Place before—and yet, tonight, I must sleep there. When I thought of it, my tongue felt dry in my throat, in spite of my wish for knowledge. Almost I would have gone down again and faced the dogs, but I did not.

I had not gone through all the rooms when the darkness fell. When it fell, I went back to the big room looking over the city and made fire. There was a place to make fire in a box with wood in it, though I do not think they cooked there. I wrapped myself in a floor-covering and slept in front of the fire—I was very tired.

Now I tell what is very strong magic. I woke in the midst of the night. When I woke, the fire had gone out and I was cold. It seemed to me that all around me there were whisperings and voices. I closed my eyes to shut them out. Some will say that I slept again, but I do not think that I slept. I could feel the spirits drawing my spirit out of my body as a fish is drawn on a line.

Why should I lie about it? I am a priest and the son of a priest. If there are spirits, as they say, in the small Dead Places near us, what spirits must there not be in that great Place of the Gods? And would not they wish to speak? After such long years? I know that I felt myself drawn as a fish is drawn on a line. I had stepped out of my body—I could see my body asleep in front of the cold fire, but it was not I. I was drawn to look out upon the city of the gods.

It should have been dark, for it was night, but it was not dark.

Everywhere there were lights—lines of light—circles and blurs of light—ten thousand torches would not have been the same. The sky itself was alight, you could barely see the stars for the glow in the sky. I thought to myself "This is strong magic" and trembled. There was a roaring in my ears like the rushing of rivers. Then my eyes grew used to the light and my ears to the sound. I knew that I was seeing the city as it had been when the gods were alive.

That was a sight indeed—yes, that was a sight: I could not have seen it in the body—my body would have died. Everywhere went the gods, on foot and in chariots—there were gods beyond number and counting and their chariots blocked the streets. They had turned night to day for their pleasure—they did not sleep with the sun. The noise of their coming and going was the noise of many waters. It was magic what they could do—it was magic what they did.

I looked out of another window—the great vines of their bridges were mended and the god-roads went East and West. Restless, restless, were the gods and always in motion! They burrowed tunnels under rivers—they flew in the air. With unbelievable tools they did giant works—no part of the earth was safe from them, for, if they wished for a thing, they summoned it from the other side of the world. And always, as they labored and rested, as they feasted and made love, there was a drum in their ears—the pulse of the giant city, beating and beating like a man's heart.

Were they happy? What is happiness to the gods? They were great, they were mighty, they were wonderful and terrible. As I looked upon them and their magic, I felt like a child—but a little more, it seemed to me, and they would pull down the moon from the sky. I saw them with wisdom beyond wisdom and knowledge beyond knowledge. And yet not all they did was well done—even I could see that—and yet their wisdom could not but grow until all was peace.

Then I saw their fate come upon them and that was terrible past speech. It came upon them as they walked the streets of their city. I have been in the fights with the Forest People—I have seen men die. But this was not like that. When gods war with gods, they use weapons we do not know. It was fire falling out of the sky and a mist that poisoned. It was the time of the Great Burning and the Destruction. They ran about like ants in the streets of their city—poor gods, poor gods! Then the towers began to fall. A few escaped—yes, a few. The legends tell it. But, even after the city had become a Dead Place, for many years the poison was still in the ground. I saw it happen, I saw the last of them die. It was darkness over the broken city and I wept.

All this, I saw. I saw it as I have told it, though not in the body. When I woke in the morning, I was hungry, but I did not think first of my hunger for my heart was perplexed and confused. I knew the reason for the Dead Places but I did not see why it had happened. It seemed to me it should not have happened, with all the magic they had. I went through the house looking for an answer. There was so much in the house I could not understand—yet I am a priest and the son of a priest. It was like being on one side of the great river, with no light to show the way.

Then I saw the dead god. He was sitting in his chair, by the window, in a room I had not entered before and, for the first moment, I thought that he was alive. Then I saw the skin on the back of his hand—it was like dry leather. The room was shut, hot and dry—no doubt that had kept him as he was. At first I was afraid to approach him—then the fear left me. He was sitting looking out over the city—he was dressed in the clothes of the gods. His age was neither young nor old—I could not tell his age. But there was wisdom in his face and great sadness. You could see that he would not have run away. He had sat at his window, watching his city die—then he himself had died. But it is better to lose one's life than one's spirit—and you could see from the face that his spirit had not been lost. I knew that, if I touched him, he would fall into dust—and yet, there was something unconquered in the face.

That is all of my story, for then I knew he was a man—I knew then that they had been men, neither gods nor demons. It is a great knowledge, hard to tell and believe. They were men—they went a dark road, but they were men. I had no fear after that—I had no fear going home, though twice I fought off the dogs and once I was hunted for two days by the Forest People. When I saw my father again, I prayed and was purified. He touched my lips and my breast, he said, "You went away a boy. You come back a man and a priest." I said, "Father, they were men! I have been in the Place of the Gods and seen it! Now slay me, if it is the law—but still I know they were men."

He looked at me out of both eyes. He said, "The law is not always the same shape—you have done what you have done. I could not have done it in my time, but you come after me. Tell!"

I told and he listened. After that, I wished to tell all the people but he showed me otherwise. He said, "Truth is a hard deer to hunt. If you eat too much truth at once, you may die of the truth. It was not idly that our fathers forbade the Dead Places." He was right—it is better the truth should come little by little. I have learned that, being a priest. Perhaps, in the old days, they ate knowledge too fast.

Nevertheless, we make a beginning. It is not for the metal alone that we go to the Dead Places now, there are the books and the writings. They are hard to learn. And the magic tools are broken—but we can look at them and wonder. At least, we make a beginning. And, when I am chief priest we shall go beyond the great river. We shall go to the Place of the Gods—the place newyork—not one man but a company. We shall look for the images of the gods and find the god ASHING and the others—the gods Lincoln and Biltmore and Moses. But they were men who built the city, not gods or demons. They were men. I remember the dead man's face. They were men who were here before us. We must build again.

Cosmic Spite

~

BY

GRACE

NICHOLS

We, the people, 'third in the world'
Feet courting the sands and mud
Of natural disasters.
After the hurricane, the floods, the famines,
The droughts and foreign debts
We chew the biblical philosophy wonderingly—
To him that hath even more will be given
To him that hath not...
But we keep on stirring rich dreams
Into the groundy porridge for our children
We keep on—the rhythm of our hard sweet lives
Despite the cosmic spite.

Sugar Plums
and Calabashes

~

BY

BRONWEN

WALLACE

On that famous poem about the night before Christmas, "the children are nestled all snug in their beds, while visions of sugar plums danced in their heads." Sugar plums are dancing in my head too, as Christmas approaches. They're dancing in a lot of women's heads, I'll bet. Well, not sugar plums exactly, but chocolates, certainly, especially the ones that still have to be purchased for dear Aunt Lillian. And the turkey of course. There it is waltzing away with the mince pie and the fruitcake. And then there's the glazed yams, and the stuffing, and the mashed potatoes, and the gravy, and the short-bread, and the jellied salads and...

All of which would be very enjoyable except that someone has to plan for all this stuff, the sugar plum fairy being stubbornly absent from most women's kitchens. Someone has to buy it and organize it and cook it. In most cases that someone will be female.

Now, I'm not saying that's all bad. I like to cook and I like to use recipes that have been handed down to me from my foremothers. I like the particularly female sense of community that comes from shar-ing recipes and cooking secrets. Mind you, I think it's pretty weird that our society is so organized that women do the majority of the food pur-chase and preparation. In most other animal species, not being able to feed yourself makes you a candidate for extinction, but then we humans have managed to set things up in such a way that even the most bizarre practices seem normal.

Women and food is one of them. And because our need for food is so basic, women's almost exclusive relationship to it has fundamental and far-reaching consequences. Women are so deeply connected with our ideas of nurturing that it appears to be simply "natural," so that men's absence from that aspect of our lives seems natural too.

Yet, like any other human activity, our practices around food are socially created. And like any other socially created practice, they are gendered. As a result, our relationship to food is complex, often contradictory and frequently dangerous.

Take women and fat, for example. Strange, isn't it, that in a society in which women are expected to be good cooks and are valued for their nurturing qualities, we are also expected to be boyishly thin. Women who obviously nurture *themselves* and who, as a result, are somewhat plump, are considered socially unattractive. A recent ad for weight-loss products said it all. It pictured a young, pretty, plump woman with the caption, "I was tired of being told I had a nice personality." Very economical, that. How it manages to insult woman's mind *and* her body in one sentence.

Taken further, this contradiction results in several deaths each year, especially among young women. Anorexia nervosa is a condition in which women (and some young men) literally starve themselves to conform to an over-exacting idea of perfect thinness. Thousands of others suffer from other forms of eating disorders which severely affect their health. The causes of these disorders are complex, but one factor is certainly a neurotic concern with body-image (and the fear of being fat) which is socially conditioned.

Indeed, the focus on body-image affects most women in our culture. Few women really *like* how we look and most of us spend huge amounts of money and time trying to get our appearance to conform to some socially acceptable—and totally unrealistic—ideal. In Victorian times it was considered unladylike to eat heartily in public. Women also laced themselves into corsets so tight they displaced their vital organs and changed the shape of their bones. We think of them now as "old-fashioned," while we drug ourselves with diet pills, starve ourselves on grapefruit and black coffee and spend hours doing bizarre exercises which have little to do with maintaining good physical health. So much for change.

What we eat—and who gets to eat it—is also socially determined. In Kingston, [Ontario], 5592 people asked for food from the Food Bank between 1985 and 1987. Of these, 50 per cent have been children and 41 per cent have been female-led families. In North America, the number of people who do not have enough to eat rises to millions. Most of these—the overwhelming majority—are women and children. As a society, we say we value children and women who care for them. But talk is cheap. Money, on the other hand, buys domed stadiums for men to play games in, not food.

Outside of North America, millions of people die daily from starvation. And here, too, the causes of their starvation are socially determined and very familiar. In many countries, land has been taken over for crops like cotton, coffee, carnations, and tobacco, which are aimed at North American markets. This leaves little good land on which to grow food. In order to survive, most people must work on the export plantation for wages *and* maintain a small plot to grow their own food as well.

In countries like Guatemala or El Salvador it is very often the women who work as low-wage workers on the plantations, forcing the men into the cities in search of other jobs. This situation, in addition to social custom, means that the women also maintain the small farm plots necessary to family survival.

In Africa, women produce approximately 80 per cent of the food grown on the continent. Many men, especially if wage-labor is not available, remain idle. A woman in Zaire may spend up to 8 hours a day in the fields, traveling up to 10 kilometres from her home. On her way back, she will carry a load of firewood weighing 30 or 40 kilos. She will also have to carry water several times a day, usually in calabashes weighing up to 20 kilos, from a source which is often 4 kilometres away.

Although they produce a large proportion of their country's food, women in Third World countries don't get to eat it. From the time they are weaned, girls generally receive less food than the boys; less food, in fact, than they need to develop normally. The result is a cycle of malnutrition in which young, undernourished mothers give birth to small babies whom they cannot adequately nurse. Complications of pregnancy and childbirth are the most common cause of death in women under 45 in developing countries. Half of these are associated with preventable undernutrition.

Of course, the practice of feeding women less is not peculiar to developing countries. In North American families too, women often eat less—or less wholesomely—than men. Sometimes they are forced to do this by husbands who control the family income; sometimes they are coerced by more subtle, but equally dangerous social pressures which urge women to be self-sacrificing.

Like women here, women in other countries are working to change their situation. In countries like Nicaragua, women are heavily involved in literacy campaigns, in the building of local health clinics, and in regaining land for co-operative use rather than export production. Projects like these are often the target of Contra attacks. Elsewhere, women are working to change a system of development aid, which, by concentrating on high technology, generally male-centered projects, often increases, or at best does not change, women's exploitation.

For the next few weeks, those of us who celebrate Christmas (and those of us who don't) will be deluged with images of feasting and plenty. And the dominant icon everywhere will be that of a mother and her child. The feasting and the plenty, in fact, are meant to convey the idea that the birth of a child is an event to celebrate.

On this planet, however, that is not always so. The politics of everyday demand that we ask certain questions. Questions like: whose child? Born where? A boy or a girl? Will she have enough to eat? Why not?

Mega-City, Mega-Problems

~

BY

ANDREA GRIMAUD

Weekends find many Canadians escaping the hustle and bustle of the city for their peaceful country cottages.

For the 50% to 75% of Third World city dwellers who live in slums there is no escape. According to the World Bank, our planet will have 23 cities of more than 10 million people by the year 2000, and 17 of these will be in poor countries.

From 1920 to 1980, the number of Third World city dwellers rose from 100 million to nearly a billion. There are now 45 cities in the Third World with popula- tions nearing, or in excess of, 3 million people. Eight have already exceeded the 10 million mark: Mexico, São Paulo, Buenos Aires, Calcutta, Bombay, Cairo, Shanghai, and Seoul.

The World Commission on Environment and Development has projected that urban growth will remain explosive until the year 2000. That means that major centres will need a 65% increase in their capacity to con- struct and manage urban infra- structure and shelter, just to hold services at their present levels.

But, at their present levels, Third World cities can't provide such basics as shelter, safe drink- ing water, health services, educa- tion, sewage, and waste disposal facilities.

In Beijing, air pollution is so bad that there are, on average, only 93 days a year when breath- ing is not a health hazard. In Lima, the road system is so poor that traffic moves an average of 6 km/h during "rush hour." And, in Cairo, 11 million Egyptians rely on sewer and water systems built for four million.

Yet, even with conditions as they are, migration continues for a variety of reasons. Many migrants are pulled by the city while being pushed from the country. The pull comes from the seeming availability of jobs and

higher wages, and the possibility of access to hospitals, electricity, and water, all of which are scarce in rural areas.

Wars, floods, drought, and other disasters force people into refugee camps which are often on the outskirts of major cities. As well, small farmers, not rich enough to modernize in order to produce export crops, often lose their land.

In the past, urban areas profited disproportionately from the social and economic policies of newly independent states. Schools, universities, hospitals, industries, and state bureaucracies were all concentrated in urban areas. This resulted in a drift of rural people to the cities.

Yet, population and environmental pressures in the cities and the countryside are inextricably linked. Through city expansion, suburbs and shantytowns spread out and swallow up adjacent farmland. And, the cities export their pollutants by air, water, and land.

Under pressure from the more politically active urban population, governments have spent more money on cities. Discrimination against rural areas is often reflected in food prices. Pressure from the urban population will often keep down official food prices which reduces income for farmers. Moving to the city seems to be their only alternative.

Their new home is most often in an illegal slum or shantytown where they face the constant threat of eviction. The land has often been left vacant by better-off citizens and industry because it is subject to flooding, landslides, and is marshy or waterlogged. Yet, the sites are chosen because they offer some advantages.

The poorer the family, the greater its need for a central location; to be sure of cheap and speedy access to jobs or sources of income.

Of Bombay's 4 million slum and sidewalk dwellers, 73% are domestic servants, luggage carriers, or casual labourers, or are self-employed as sellers of fruit and vegetables, flowers, ice-cream, and other items. Only 1.7% are unemployed.

Even the poor are not equally poor. In squatter settlements, social class division steps in. Those with the most energy, and best health and skills live in the best locations. The less able ones live farther out. But, all are equally vulnerable to disaster. Fires and explosions occur most frequently in poor areas. Houses which have little or no foundation, leaky roofing, and flimsy walls collapse easily.

Rio de Janeiro, with more than 10 million residents, is constantly subjected to floods due to intense rainfall and its geographic location in a humid tropical region. Disasters of varying degrees occur nearly every year.

The emergence of Rio's hillside shanties began at the turn of the century. Since the '50s, more than 400 shanty communities have emerged. As unsanitary as the original, they are yet farther away from work markets.

Rio de Janeiro is but one example of overcrowding. Buenos Aires has more than 10 million inhabitants in its metropolitan area; Lima, nearly 7 million, Bogota, more than 6 million. They all face similar problems.

With an economic crisis which has affected the whole of Latin America, there is less and less money to invest in the infrastructure. Transportation, pollution, water supply, and housing are all major issues, and shortcomings in these areas are felt first and most critically by the poor.

As it becomes harder to earn an honest living, crime has become a growth industry. It's estimated that one million people in Rio de Janeiro earn their living from crime or its prevention. They include not only kidnappers, murderers, and drug traffickers, but also lookouts, sellers of stolen goods, and private security guards hired to prevent assaults and break-ins. Members of the military police, who safeguard the streets, are regularly found to be in league with criminal gangs.

The situation is made worse by a shortage of effective city officials. City halls across Asia, Africa, and Latin America tend to lack one essential resource— skilled people. Local officials who show any ability are quickly recruited by the central government or private sector. Yet, even with skilled people at the higher levels of government, legislation on such issues as pollution in developing countries is often vague or nonexistent.

Many national and urban governments lack the money, or the will, to make a long-term commitment to their natural resources. Population, such as exists in large cities, is one of the major causes of water contamination. The water on which cities rely is often polluted by enormous amounts of human waste, sometimes channeled, untreated, into lakes and rivers.

There are two types of water resources used for urban water supply, surface water (rivers and lakes) and groundwater (underground wells and springs).

Scores of cities have damaged their neighbouring water bodies. Over the last 50 years, cities have increasingly used groundwater to compensate for the gradual loss of surface water.

Many of the 20 large cities in the developing world pump water from the ground to meet the people's needs: Mexico City, Bangkok, Calcutta, Manila, Jakarta, São Paulo, Buenos Aires, Beijing, and Shanghai. Water stored under the ground in aquifers is often abundant and of high quality. And, it is much less likely to be contaminated as it is fairly well protected by layers of sediment and rock.

Yet, polluted water can find its way into underground reservoirs. This invisible pollution is impossible to clean up. Buenos Aires' groundwater reserves, which account for 40% of its drinking water, are at risk of contamination by salt water. As a result of unrestricted pumping, sea water has seeped into the region's largest reservoir, the Puelche aquifer. And, a number of wells have closed.

Most Third World cities are overpumping their aquifers beyond their renewal possibilities. They are gradually drying up.

Producing cheap exports is no help to the environment either. Gold extraction operations and waste from coffee production and the tanning industry are ruining the freshwater environment in South America. Heavy industry in such cities as Cairo, Calcutta, and Mexico City is doing much the same.

Many cities have dropped their standards of water quality, allowing consumption of water which would not normally be considered safe. There is evidence that waterborne sicknesses such as diarrhea, hepatitis, and cholera have become commonplace in many cities.

It's estimated that 200 million urban dwellers in the Third World lack the benefits of safe running water. And, water resource management in developing cities is not getting better.

Recently, there has been a widespread decrease in funds for urban water supply purposes. The World Bank's financing of water projects in Latin America in relation to total funding has decreased from 7.2% in 1976-'80 to 3.9% in 1986-'89. However, many international lending institutions, including the World Bank and the Inter-American Development Bank, have said that environmental sustainability is a key element in the projects they fund.

To the detriment of cities, many First World development agencies continue to focus on rural problems such as deforestation, agriculture, and climate change. They see cities as having all the benefits of development: sewers, roads, electricity, and hospitals. Therefore, for 10 to 20 years cities have not received their due.

As long as this trend continues, the World Bank, United Nations, and other development agencies recommend lower cost solutions such as:

- better municipal management,
- controlling waste disposal,
- privatizing municipal services,
- reducing consumption by such means as raising the price of pollution-causing gasoline,
- community development,
- loosening of city regulations which often slow local development and increase costs.

Heavy subsidies are needed to bring housing within the reach of low-income families. In a small way, governments have attempted rehousing of the poor. But, as it is seen as a welfare activity, it gets low priority.

When housing is scarce even for the better-off, the poor end up selling or sub-letting their homes to wealthier families. They then return to squatter settlements with their profits. For them, living in a decent house is no substitute for badly needed food or medical care.

What constitutes a decent house, according to developed world standards, is often unacceptable to Third World citizens. The majority of the world's people live in houses of mud, timber, thatch, and stone—natural materials suited to their environment. Yet, to the people's dissatisfaction, development efforts have often provided Western-style houses made of concrete. With the era of cheap cement coming to an end, a return to traditional materials and construction techniques offers a good alternative.

Many development agencies argue that wealthy nations must increase aid budgets and write off loans in order to allow Third World governments to fund local initiatives. Others believe that to do so would signal to Third World leaders that they could resume borrowing and that past acts of inhumanity were being condoned.

In collaboration with the World Bank, European and North American commercial banks invested in megaprojects in primarily dictator-run countries in the '70s.

One trillion dollars was used to erect power plants, steel mills, nuclear reactors, and chemical factories which displaced millions of people and employed far less. With the high interest rates of the '80s, which caused debt payments to spiral, many projects died leaving more unemployment and wasted land.

Land will always remain a very sensitive, political issue. Governments are reluctant to hand over such a valuable commodity to those who have no influence. Studies have found that they should think more of providing incentives which would keep people in the rural areas, such as the development of large villages or small towns to which people could move, but without moving very far.

Improvement of health services can help reduce the need that many parents feel to have large families. And education, particularly of women, has an important role to play. Many studies show that the longer women have been in school, the more likely they are to work outside the home and not only have, but lose, fewer children.

Solving the problems of Third World countries, their cities, and their people may never be fully attained. In order to do so, all governments must learn to take into account the special needs of those who live at the bottom of the social scale.

DATA BANK

Population of Detroit in 1955: 2 million
in 1992: 1 million

Approximate number of cities in the world with populations greater than 100 000 in the year 1800: 50
in the year 1970: 1500

Number of cities in the world with 10 million or more people by 2000: 23
Number of these cities that will be in the developing world: 17

Percentage of Calcutta's residents that live in slums: 66

Percentage of families in Manila that have access to the city's water system: 10

Percentage of time public buses in Kinshasa, Zaire, are out of service for repairs: 90

Cost of musical entertainment for mayors of Third World cities gathered in Montreal in 1991 to discuss problems of urban areas in developing countries: $600 000

Number of new people settling in the Third World's cities each day: 160 000

Average number of days in a year when the air in Beijing is so polluted that breathing becomes a health hazard: 272

Percentage of China's population that lived in cities in 1965: 18
in 1989: 53

Top
of the
Food Chain

~

BY

T. CORAGHESSAN

BOYLE

The thing was, we had a little problem with the insect vector there, and believe me, your tamer stuff, your malathion and pyrethrum and the rest of the so-called environmentally safe products, didn't begin to make a dent in it, not a dent. I mean it was utterly useless—we might as well have been spraying Chanel No. 5 for all the good it did. And you've got to realize these people were literally covered with insects day and night—and the fact that they hardly wore any clothes just compounded the problem. Picture if you can, gentlemen, a naked little two-year-old boy so black with flies and mosquitoes it looks like he's wearing long johns, or the young mother so racked with the malarial shakes she can't even lift a Diet Coke to her lips—it was pathetic, just pathetic, like something out of the Dark Ages...Well, anyway, the decision was made to go with DDT. In the short term. Just to get the situation under control, you understand.

Yes, that's right. Senator, *DDT*: dichlorodiphenyltrichloroethane.

Yes, I'm well aware of that fact, sir. But just because we banned it domestically, under pressure from the bird-watching contingent and the hopheads down at the EPA, it doesn't necessarily follow that the rest of the world—especially the developing world—was about to jump on the bandwagon. And that's the key word here, Senator, "developing." You've got to realize this is Borneo we're talking about here, not Port Townsend or Enumclaw. These people don't know from square one about sanitation, disease control, pest eradication—or even personal hygiene, if you want to come right down to it. It rains 120 inches a year, minimum. They dig up roots in the jungle. They've still got headhunters along the Rajang River, for God's sake.

And please don't forget they *asked* us to come in there, practically begged us—and not only the World Health Organization but the Sultan of Brunei and the government in Sarawak too. We did what we could to accommodate them and reach our objective in the shortest period of time and by the most direct and effective means. We went to the air. Obviously. And no one could have foreseen the consequences, no one, not even if we'd gone out and generated a hundred environmental impact statements—it was just one of those things, a freak occurrence, and there's no defense against that. Not that I know of, anyway...

Caterpillars? Yes, Senator, that's correct. That was the first sign: caterpillars.

But let me backtrack a minute here. You see, out in the bush they have these roofs made of thatched palm leaves—you'll see them in the towns too, even in Bintulu or Brunei—and they're really pretty effective, you'd be surprised. A hundred and twenty inches of rain, they've got to figure a way to keep it out of the hut, and for centuries, this was it. Palm leaves. Well, it was about a month after we sprayed for the final time and I'm sitting at my desk in the trailer thinking about the drainage project at Kuching, enjoying the fact that for the first time in maybe a year I'm not smearing mosquitoes all over the back of my neck, when there's a knock at the door. It's this elderly gentleman, tattooed from head to toe, dressed only in a pair of running shorts—they love those shorts, by the way, the shiny material and the tight machine-stitching, the whole country, men and women both, they can't get enough of them...Anyway, he's the headman of the local village and he's very excited, something about the roofs—*atap,* they call them. That's all he can say, *atap, atap,* over and over again.

It's raining, of course. It's always raining. So I shrug into my rain slicker, start up the 4 x 4, and go have a look. Sure enough, all the atap roofs are collapsing, not only in his village but throughout the target area. The people are all huddled there in their running shorts, looking pretty miserable, and one after another the roofs keep falling in, it's bewildering, and gradually I realize the headman's diatribe has begun to feature a new term I was unfamiliar with at the time—the word for caterpillar, as it turns out, in the Iban dialect. But who was to make the connection between three passes with the crop duster and all these staved-in roofs?

Our people finally sorted it out a couple weeks later. The chemical, which, by the way, cut down the number of mosquitoes exponentially, had the unfortunate side effect of killing off this little wasp—I've got the scientific name for it somewhere in my report here, if you're interested—that preyed on a type of caterpillar that in turn ate palm leaves. Well, with the wasps gone, the caterpillars hatched out with nothing to keep them in check and chewed the roofs to pieces, which was unfortunate, we admit it, and we had a real cost overrun on replacing those roofs with tin...but the people were happier, I think, in the long run, because, let's face it, no matter how tightly you weave those palm leaves, they're just not going to keep the water out like tin. Of course, nothing's perfect, and we had a lot of complaints about the rain drumming on the panels, people unable to sleep, and what have you...

Yes, sir, that's correct—the flies were next.

Well, you've got to understand the magnitude of the fly problem in Borneo, there's nothing like it here to compare it with, except maybe a garbage strike in New York. Every minute of every day you've got flies everywhere, up your nose, in your mouth, your ears, your eyes, flies in your rice, your Coke, your Singapore sling, and your gin rickey. It's enough to drive you to distraction, not to mention the diseases these things carry, from dysentery to typhoid to cholera and back round the loop again. And once the mosquito population was down, the flies seemed to breed up to fill in the gap—Borneo wouldn't be Borneo without some damned insect blackening the air.

Of course, this was before our people had tracked down the problem with the caterpillars and the wasps and all of that, and so we figured we'd had a big success with the mosquitoes, why not a series of ground sweeps, mount a fogger in the back of a Suzuki Brat, and sanitize the huts, not to mention the open sewers, which as you know are nothing but a breeding ground for flies, chiggers, and biting insects of every sort—at least it was an error of commission rather than omission. At least we were trying.

I watched the flies go down myself. One day they were so thick in the trailer I couldn't even *find* my paperwork, let alone attempt to get through it, and the next thing they were collecting on the windows, bumbling around like they were drunk. A day later they were gone. Just like that. From a million flies in the trailer to none...

Well, no one could have foreseen that, Senator.

The geckos ate the flies, yes. You're all familiar with geckos, I assume, gentlemen? These are the lizards you've seen during your trips to Hawaii, very colorful, patrolling the houses for roaches and flies, almost like pets, but of course they're wild animals, never lose sight of that, and just about as unsanitary as anything I can think of, except maybe flies.

Yes, well don't forget, sir, we're viewing this with twenty-twenty hindsight, but at the time no one gave a thought to geckos or what they ate—they were just another fact of life in the tropics. Mosquitoes, lizards, scorpions, leeches—you name it, they've got it. When the flies began piling up on the windowsills like drift, naturally the geckos feasted on them, stuffing themselves till they looked like sausages crawling up the walls. Whereas before they moved so fast you could never be sure you'd seen them, now they waddled across the floor, laid around in the corners, clung to the air vents like magnets—and even then no one paid much attention to them till they started turning belly-up in the streets. Believe me, we confirmed a lot of things there about the buildup of these products as you move up the food chain and the efficacy—or lack thereof—of certain methods, no doubt about that...

The cats? That's where it got sticky, really sticky. You see, nobody really lost any sleep over a pile of dead lizards—though we did tests routinely and the tests confirmed what we'd expected, that is, the product had been concentrated in the geckos because of the number of contaminated flies they consumed. But lizards are one thing and cats are another. These people really have an affection for their cats—no house, no hut, no matter how primitive, is without at least a couple of them. Mangy-looking things, long-legged and scrawny, maybe, not at all the sort of animal you'd see here, but there it was: they loved their cats. Because the cats were functional, you understand—without them, the place would have been swimming in rodents inside of a week.

You're right there, Senator, yes—that's exactly what happened.

You see, the cats had a field day with these feeble geckos—you can

imagine, if any of you have ever owned a cat, the kind of joy these animals must have experienced to see their nemesis, this ultra-quick
lizard, and it's just barely creeping across the floor like a bug. Well, to
make a long story short, the cats ate up every dead and dying gecko in
the country, from snout to tail, and then the cats began to die…which
to my mind would have been no great loss if it wasn't for the rats.
Suddenly there were rats everywhere—you couldn't drive down the
street without running over half a dozen of them at a time. They
fouled the grain supplies, fell in the wells and died, bit infants as they
slept in their cradles. But that wasn't the worst, not by a long shot. No,
things really went down the tube after that. Within the month we were
getting scattered reports of bubonic plague, and of course we tracked
them all down and made sure the people got antibiotics, but still we
lost a few and the rats kept coming…

It was my plan, yes. I was brainstorming one night, rats scuttling
all over the trailer like something out of a cheap horror film, the villagers in a panic over the threat of the plague and the stream of non-
stop hysterical reports from the interior—people were turning black,
swelling up and bursting, that sort of thing—well, as I say, I came up
with a plan, a stopgap, not perfect, not cheap, but at this juncture, I'm
sure you'll agree, something had to be done.

We wound up going as far as Australia for some of the cats, clean-
ing out the S.P.C.A. facilities and what have you, though we rounded
most of them up in Indonesia and Singapore—approximately 14,000 in
all. And yes, it cost us—cost us up-front purchase money and aircraft
fuel and pilots' overtime and all the rest of it—but we really felt there
was no alternative. It was like all nature had turned against us.

And yet, all things considered, we made a lot of friends for the
U.S.A. the day we dropped those cats, and you should have seen them,
gentlemen, the little parachutes and harnesses we'd tricked up,
14,000 of them, cats in every color of the rainbow, cats with one ear,
no ears, half a tail, three-legged cats, cats that could have taken pride
of show in Springfield, Massachusetts, and all of them twirling down
out of the sky like great big oversized snowflakes…

It was something. It was really something.

Of course, you've seen all the reports. There were other factors we
hadn't counted on, adverse conditions in the paddies and manioc
fields—we don't to this day know what predatory species were inad-
vertently killed off by the initial sprayings, it's just a mystery—but the

weevils and whatnot took a pretty heavy toll on the crops that year, and by the time we dropped the cats, well—the people were pretty hungry, and I suppose it was inevitable that we lost a good proportion of them right then and there. But we've got a CARE program going there now and something hit the rat population—we still don't know what, a virus, we think—and the geckos, they tell me, are making a comeback.

So what I'm saying is it could be worse, and to every cloud a silver lining, wouldn't you agree, gentlemen?

The Discovery of Poverty

~

BY

WOLFGANG SACHS

I could have kicked myself afterwards. Yet my remark had seemed the most natural thing on earth at the time. It was six months after Mexico City's catastrophic earthquake in 1985 and I had spent the whole day walking around Tepito, a dilapidated quarter inhabited by ordinary people but threatened by land speculators. I had expected ruins and resignation, decay and squalor, but the visit had made me think again: there was a proud neighbourly spirit, vigorous building activity and a flourishing shadow economy.

But at the end of the day the remark slipped out: "It's all very well but, when it comes down to it, these people are still terribly poor." Promptly, one of my companions stiffened: *"No somos pobres, somos Tepitanos!"* (We are not poor people, we are Tepitans). What a reprimand! Why had I made such an offensive remark? I had to admit to myself in embarrassment that, quite involuntarily, the clichés of development philosophy had triggered my reaction.

Inventing the low-income bracket

"Poverty" on a global scale was discovered after the Second World War; before 1940 it was not an issue. In one of the first World Bank reports, dating from 1948-9, the "nature of the problem" is outlined: "Both the need and potential for development are plainly revealed by a single set of statistics. According to the UN Bureau of Statistics, average income per head in the United States in 1947 was over $1400, and in another 14 countries ranged between $400 and $900. For more than half of the world's population, however, the average income was less—and sometimes much less—than $100 per person. The magnitude of this discrepancy demonstrates not only the urgent need to raise living standards in the underdeveloped countries, but also the enormous possibilities to do just this..."

Whenever "poverty" was mentioned at all in the documents of the 1940s and 1950s, it took the form of a statistical measurement of per-capita income whose significance rested on the fact that it lay ridiculously far below the US standard.

When size of income is thought to indicate social perfection, as it does in the economic model of society, one is inclined to interpret any other society which does not follow that model as "low-income." This way, the perception of poverty on a global scale was nothing more than the result of a comparative statistical operation, the first of which was carried out only in 1940 by the economist Colin Clark. As soon as the scale of incomes had been established, order reigned on a confused globe: horizontally, such different worlds as those of the Zapotec people of Mexico, the Tuareg of North Africa and the Rajasthani of India could be classed together, whilst a vertical comparison to the "rich" nations demanded relegating them to a position of almost immeasurable inferiority.

In this way "poverty" was used to define whole peoples, not according to what they are and want to be, but according to what they lack and are expected to become. Economic disdain had thus taken the place of colonial contempt.

Moreover, this conceptual operation provided a justification for intervention: wherever low income is the problem, the only answer can be "economic development." There was no mention of the idea that poverty might also result from oppression and thus demand liberation. Or that a culture of sufficiency might be essential for long-term survival. Or even less that a culture might direct its energies towards spheres other than the economic.

No, as it was in the industrial nations so it was to be in all the others: poverty was diagnosed as a lack of spending power crying to be banished through economic growth. Under the banner of "poverty" the enforced reorganization of many societies into money economies was subsequently conducted like a moral crusade. Who could be against it?

Descent to the biological minimum

Towards the end of the 1960s, when it was no longer possible to close one's eyes to the fact that "economic development" was patently failing to help most people achieve a higher standard of living, a new conception of "poverty" was required. "We should strive," McNamara of the World Bank stated in 1973, "to eradicate absolute poverty by the end of the century. That means in practice the elimination of malnutrition and illiteracy, the reduction of infant mortality and the raising of life-expectancy standards to those of the developed nations."

Whoever lived below an externally defined minimum standard was declared "absolutely poor"; the yardstick of per-capita income was thrown onto the trash heap of development concepts. Two shifts in the focus of the international discussion of poverty were responsible for this. On the one hand, attention switched to yawning social gulfs within societies, which had been completely blurred by national averages. On the other, income revealed itself to be a rather blunt indicator of the actual living conditions of those not fully integrated into a money economy.

These new efforts to understand poverty in terms of quality of life emerged out of disappointment at the results of the stimulation of growth, but they brought their own form of reductionism. Since the first attempts in England at the turn of the century, the calculation of an absolute poverty line has been based on nutrition: the absolute poor are those whose intake of foods does not exceed a certain minimum of calories.

The trouble with such definitions is that they reduce the living reality of hundreds of millions of people to an animalistic description. In an attempt to find an objective and meaningful criterion, the ground was cleared for a conception of reality which reduces the rich variety of what people might hope and struggle for to one bare piece of data about survival. Can a lower common denominator be imagined?

No wonder the measures taken in response—ranging from deliveries of grain to people who eat rice to literacy campaigns in regions where the written word is altogether uncommon—have all too often been insensitive and shown no regard for people's self-esteem.

Reducing lifeworlds to calorie levels, to be sure, makes the international administration of development aid a lot easier. It allows a neat classification of the clientele (without which world-wide strategies would be pointless) and it serves as permanent proof of a state of global emergency (without which doubt might be cast on the legitimacy of some development agencies).

This readjusted idea of poverty enabled the development paradigm to be rescued at the beginning of the 1970s. In its official version, the fulfilment of basic needs strictly called for economic growth, or at least growth "with redistribution." The link to the previous decade's dogma of growth was thus established.

Poor is not necessarily poor

Binary divisions, such as healthy/ill, normal/abnormal or, more pertinently, rich/poor, are like steamrollers of the mind; they level a multiform world, completely flattening anything which does not fit. The stereotyped talk of "poverty" has disfigured the different, indeed contrasting, forms of poverty beyond recognition. It fails to distinguish, for example, between frugality, destitution, and scarcity.

Frugality is a mark of cultures free from the frenzy of accumulation. In these, the necessities of everyday life are mostly won from subsistence production with only the smaller part being purchased on the market. To our eyes, people have rather meagre possessions; maybe the hut and some pots and the Sunday costume, with money playing only a marginal role. Instead, everyone usually has access to fields, rivers and woods, while kinship and community duties guarantee services which elsewhere must be paid for in hard cash. Despite being in the "low-income bracket," nobody goes hungry. What is more, large surpluses are often spent on jewellery, celebrations or grandiose buildings. In a traditional Mexican village, for example, the private accumulation of wealth results in social ostracism—prestige is gained precisely by spending even small profits on good deeds for the community. Here is a way of life maintained by a culture which recognizes and cultivates a state of sufficiency; it only turns into demeaning "poverty" when pressurized by an accumulating society.

Destitution, on the other hand, becomes rampant as soon as frugality is deprived of its foundation. Along with community ties, land, forest and water are the most important prerequisites for subsistence without money.

As soon as they are taken away or destroyed, destitution lurks. Again and again, peasants, nomads and tribals have fallen into misery after being driven from their land, savannahs and forests. Indeed the first state policies on poverty, in sixteenth-century Europe, were a response to the sudden appearance of vagabonds and mendicancy provoked by enclosures of the land— it had traditionally been the task of communities to provide for widows and orphans, the classical cases of unmaintained poor people.

Scarcity derives from modernized poverty. It affects mostly urban groups caught up in the money economy as workers and consumers whose spending power is so low that they fall by the wayside. Not only does their predicament make them vulnerable to the whims of the market, but they also live in a situation where money assumes an ever-increasing importance. Their capacity to achieve through their own efforts gradually fades, while at the same time their desires, fuelled by glimpses of high society, spiral towards infinity; this scissor-like effect of want is what characterizes modern poverty. Commodity-based poverty, still described as "the social question" in the nineteenth century, led to the welfare state and its income and employment policy after the world economic crisis of 1929. Precisely this view of poverty, influenced by Keynes and the New Deal, shaped the development ideas of the post-war era.

More frugality, less destitution

Up until the present day, development politicians have viewed "poverty" as the problem and "growth" as the solution. They have not yet admitted that they have been largely working with a concept of poverty fashioned by the experience of commodity-based need in the Northern hemisphere. With the less well-off *homo oeconomicus* in mind, they have encouraged growth— and often produced destitution— by bringing multifarious cultures of frugality to ruin. For the culture of growth can only be erected on the ruins of frugality; and so destitution and dependence on commodities are its price.

Is it not time after 40 years to draw a conclusion? Whoever wishes to banish poverty must build on efficiency; a cautious handling of growth is the most important way of fighting poverty.

It seems my friend from Tepito knew of this when he refused to be labelled "poor." His honour was at stake, his pride

too; he clung to his Tepito form : destitution or never-ending
of sufficiency, perhaps sensing : scarcity of money.
that without it there loomed only :

It's Time to Rethink how we Divide the World

~

BY

PAUL

HARRISON

ince the Iron Curtain collapsed, we have been in a morass about what we used to call the Third World. If we do not get the terms right, there is a good chance we will not get the policies right either.

Preparing a third edition of *Inside the Third World* (published by Penguin), brought the dilemma to my doorstep. How could I keep the title? Thirteen years after the first edition, the First, rich capitalist world was more firmly in place than ever; but the Second, Communist world had all but vanished. The middle term had disappeared: how could the third still be Third?

But the alternatives had drawbacks, too. "Developing countries"—the bureaucrat's term—is long, boring, and value-laden. Developed countries are developing, too—towards God knows what.

The current favorite of the activists is South. North and South are short, staccato words. They have a nice ring of conflict and inequality about them. But they, too, are inexact. The South is not all poor, nor is the North all rich. Large parts of China, Turkey and North Korea lie to the north of Washington. And in the far south, beyond the heat-cursed tropics, is another band of relative prosperity stretching from Uruguay through white South Africa to Australasia.

The worst thing about these options is that they divide countries all too neatly into two. Had the Iron Curtain fallen 30 years ago, that might have been a good way of looking at the world. Today it is a good way of blinding oneself to the world.

The globe is no longer polarized into have-alls and

have-nothings. There is a sizeable middle class of 29 countries where average incomes per person—in terms of comparable purchasing power—range between $4000 to $8500. These include oil exporters like Syria and Iran, and emerging industrialists like Thailand or South Korea.

Singapore, Hong Kong and Israel are now classed as high-income countries by the World Bank, along with Kuwait and the United Arab Emirates. Yet all of these are still listed as "developing countries" by the Western aid club, the Development Assistance Committee. The DAC list has barely changed since it was first drawn up 30 years ago. Only Spain and Portugal have dropped off it, at their own request.

The two-world terminology not only oversimplifies. It serves as a device to conceal how much aid is going to those who don't need it, so aid can be determined by power politics or self-interest, rather than by social and environmental needs. That is surely not why Western taxpayers support aid budgets.

Aid currently bears little relation to need. Lower-middle income countries get more than twice as much aid per person as low income countries—though their average income is three times higher. Even upper-middle income countries, where average incomes are nearly seven times higher than in low income countries, get nearly 60 per cent more aid per person.

Some of the disparities are grotesque. Each Israeli got $351 of aid in 1990-91—10 times more than the average person from dirt-poor Madagascar with its disappearing soils and species.

Each Malaysian gets 50 per cent more aid than the average disaster-prone Bangladeshi, though Malaysian personal incomes are 10 times higher. In 1991, North Africa and the Middle East got 35 per cent more total aid than south Asia, with four times the population and a quarter of the income per person.

There are diplomatic consequences, too. In world environmental negotiations about climate change or deforestation, the better-off in the South can huddle together with the really poor and pretend that they are all in the same frail boat, swamped by the rich wake of the North.

Countries such as Malaysia (purchasing power per person $5900) can divert attention from their own reckless deforestation and rocketing car use to "Northern overconsumption."

Arab oil-exporters, which have carbon dioxide emissions per person higher than in Europe, can make common cause with India and China in resisting Northern calls for restraint.

We need more precision, but there is a Babel of terms in use. The World Bank splits nations up into a class system of lower income, middle income and high income countries—LICs, MICs and HICs, for short—with the middle subdivided into upper and lower middle, UMICs and LMICs. There are the newly industrializing countries (NICs), the DAEs (dynamic Asian economies) and the LDCs (least developed countries).

Economics aren't everything. The United Nations Development Program, more concerned with the quality of life, divides countries up into high, medium and low human development. UNICEF splits them down by child death rates.

Among all these trees, we need a perspective view of the woods. Precision is desperately needed in the right context. We need to target aid much more finely to the poor and to the sensitive environments that need it most. For everyday use we need something that trips off the tongue, without too many sub-tleties. But also something with meaning, something that helps to make ethical and political sense of the world, something that can serve as a better guide to action.

When you chart how countries are distributed by purchasing power, you notice a definite clumping.

There are three peaks—39 countries below $2000, then 28 countries between $4000 and $8000, and 17 over $14 000. The other 20 are ranged in the two gaps. So we can still talk of three worlds, but in a different and more useful sense.

There is a rich First World of about 900 million people, with incomes above $6000 a year. This group should be providing the aid and reducing its own environmental impact radically.

Then there is a middle-income Second World of around one billion people, with incomes between $1500 and $6000. This second group needs trade concessions and loans and technological co-operation to help it leap over Western environmental errors. But does it need aid?

The Third World has the poorest countries, with dollar incomes below $1500. This world numbers almost 3.5 billion—indeed 2.9 billion of them are clustered below $500 a year income. Here poverty and rapid population growth combine to damage the local environment—not other people's. There is a strong case for targeting all aid on this group, focusing it on human resource development and environmental conservation.

We still need the term Third World. Redefined, it can help to focus our thoughts and assistance on those who need them most.

Why Asia's Tigers Burn So Bright

BY

JOHN STACKHOUSE

alaysia's long climb from poverty is a tale of Golden Hope.

In a little more than a decade, the country's largest rubber and palm-oil producer, Golden Hope Plantations, has transformed itself from a simple estate manager into a diversified multinational corporation with 25 000 employees.

The Malaysian managers did it by making decisions their former British bosses, who sold the firm in 1982, never considered: they poured millions of dollars into research, developed a new type of rubber tree twice as productive as its competition in Indonesia and built factories to manufacture the finished products, from rubber boots to palm-oil margarine to Ikea furniture parts.

"We always have to stay one step ahead," says Radzuan Abdul Rahman, a senior executive of the company.

The drive to stay ahead has been so successful that by the end of the decade Golden Hope expects to buy much of its raw rubber from Vietnam, Bangladesh and possibly Nigeria, and process it in Malaysia, just as its colonial owners once processed Malaysian rubber in Britain.

Golden Hope's success mirrors that of its country and its region. Through planning, investment, competition and an obsession with growth, Malaysia and Southeast Asia as a whole have tripled their incomes in one generation. In doing so, they have neared the end of their long march from poverty.

With the notable exception of the Philippines (which is catching up), Southeast Asia's economy has taken off. For five years running it has been the world's fastest-growing region, outpacing gains in the global economy by 6 to 1.

Over the long haul, it has outstripped other developing regions by an even greater margin. Between 1965 and 1990, Southeast Asia's economy grew 300 per cent, South America's by only 40 per cent and South Asia by 30 per cent. Sub-Saharan Africa barely managed to hold its ground.

Now many countries in the region look to the future and worry that the next step in their development may be more difficult. Nevertheless, their past accomplishment—the rapid transformation of a crowded rural backwater into an economic powerhouse—is being studied closely by many new kids on the trading block.

The change has been epochal for consumers, investors, managers, entrepreneurs, labourers, farmers, teachers, students—and all those who try to lead them. Once dependent on foreign markets, the region's economy can now turn to its own consumers. Malaysians last year bought $10-billion worth of laptop computers, televisions, VCRs, stereos and other consumer electronics, most of them made at home.

In Thailand, auto sales grew 30 per cent in 1993—Thais now buy more pickup trucks than any country outside the United States—while department-store sales jumped 14.5 per cent. The 600 000 consumers entering the 25-to-44 age bracket every year in Thailand are the reason Toyota plans to build its second-largest plant in the world there.

The wealth has spread to the countryside, too. On the main street of Nang Rong, a formerly impoverished town in Northeast Thailand, Song Sneg Supyon, a farmer who sold his land to open a store, now sells 20 motorized plows a month, each for 45 000 baht ($2500). A big new private hospital is opening down the road, and it's hard to find a farmhouse in the surrounding countryside that does not have the modern essentials of rural Thailand: a television, electric rice cooker, Panasonic tape player and electric fan.

Lack of skills? In 1991, Thai students scored better on international algebra tests than students from Sweden and only slightly behind those from Ontario, England and New Zealand.

Lack of capital? Financial markets are thriving. When Malaysian developer YTL Corp. needed $1-billion recently for a few power projects, it raised the money in one shot in Kuala Lumpur.

But for all the cellular phones, cars and condos, Southeast Asia's greatest achievement has been the huge strides it has made against poverty. In 1971, 68 million Indonesians were classified as poor; in 1990, despite a 50-per cent increase in population, the number had dropped to 18 million. In Malaysia, infants born today are five times more likely to survive than they were in 1960. In Thailand, people can expect to live to 66—14 years longer than their parents, and, sadly, also 14 years longer than the average African.

How did Asia's new little dragons achieve such phenomenal growth when so many other developing regions stagnated? The reasons are as varied as the countries themselves, but the most frequently cited ones are:

Government
More than anything else, Southeast Asia benefits from governments that are obsessed— sometimes at any cost—with economic development. Social investments are considered a sacred trust, inflation is seen as anathema, foreign investment and exports are national priorities.

"The governments in this region have all the right hangups," says Chalongphob Sussangkarn, an economist at the Thailand Development Research Institute.

When Indonesia's oil boom ended in the mid-1980s and the government faced a budget crisis, President Suharto refused to cut education and health care; economic subsidies, megaprojects and defence spending went out the window first.

Thailand's and Indonesia's governments both work under balanced-budget legislation.

To keep bureaucrats honest, Indonesia grades its local officials on the basis of development targets for agricultural output, school enrolment, sanitation and population growth.

The region's governments are even more obsessed with rising prices. They learned in the 1960s that inflation (it hit 600 per cent in Indonesia) is the single biggest obstacle to savings and investment, and have fought it vigorously. From 1980 to 1991, Thailand's inflation rate averaged 3.7 per cent annually, while Malaysia's was 1.7 per cent. Indonesia was the region's delinquent, allowing inflation to run at an average of 8.5 per cent a year for the entire decade.

Economists even have a good word to say about the endemic corruption. "In Africa, government people steal a lot and put it in Swiss bank accounts," says Mohamad Sadli, a former Indonesian cabinet minister. "Here, they take the money and start their own businesses. Everyone wants to be in business here."

Education

When Indonesia reaped generous oil windfalls in the 1970s, it committed 50 per cent of the money to development projects: primary schools, rural health posts and irrigation canals led the way. Thailand and Malaysia devote one-fifth of all public spending to education, and most of that goes to primary and secondary schools, not universities.

More money doesn't guarantee better education, but it means at least this: the pupil-teacher ratio is 24 to 1 in Indonesia, 21 in Malaysia and 18 in Thailand. It is 41 in Pakistan, 46 in India and 60 in Bangladesh.

Infrastructure

Highways, dams, airports, canals—these are the dirty words of development today. Southeast Asia invested in them anyway.

To win back peasants from the communist insurgencies of the 1960s, Thailand, Malaysia and Indonesia poured millions into rural infrastructure, paving roads and stringing hydro lines into every village. During the 1980s, Thailand continued the program by doubling the length of its provincial highways and bringing electricity to about 36 700 villages. Today, 95 per cent of Thailand's villages have electricity and 74 per cent have paved farm-to-market roads.

Exports

In 1970, exports accounted for 15 per cent of Thailand's economic output. In 1993, it was 40 per cent. The difference is worth about 100 000 new jobs a year.

Attracting export industries involves more than opening doors. It requires governments to make politically painful decisions: to divest state trading corporations, encourage foreign investment, reduce tariffs.

One decision that Thailand, Malaysia and Indonesia made was to tax consumers instead of traders. None of these new economic tigers earns more than 20 per cent of its government revenue from taxes and tariffs on international trade. India, Pakistan and the Philippines, on the other hand, earn no less than 25 per cent of their revenue from that source.

Competition

"We mistakenly tried to protect our industries from foreign

competition," Philippine President Fidel Ramos said recently. "We mistakenly equated nationalism with economic self-sufficiency."

Anyone who disagrees with Mr. Ramos might want to compare Thailand's protected textiles industry with its unprotected garments industry. Behind the wall of a 60-per-cent tariff, the textiles industry has not grown in years. Its sales of $1.4 billion last year did not put it among the world's top producers, and its payroll of 300 000 has barely budged since the 1970s.

Facing intense competition from abroad, the Thai garment industry started from scratch in 1980; in one decade it created nearly one million jobs and $5.1-billion worth of exports, ranking it ninth in the world. Mr. Ramos decided to follow the Thai garment path, and last year the Philippine's exports rose 16 per cent. Leading the way: electronics and garments.

Adding Value

Southeast Asia's governments have pushed industry to climb the value-added ladder, as Golden Hope did in shifting from rubber extraction to rubber processing.

In the 1980s, Thailand shifted its export focus from rice— Vietnam and India can take care of that—to canned and frozen food, and to specialty fruits, which are now worth double its rice exports. Not to be outdone, Singapore believes much of its future lies in biotechnology, and it is investing heavily in research into cellular and molecular biology.

Southeast Asia also made the shift in primary resources. By banning log exports in 1985, Indonesia now produces 7 per cent of the world's plywood. Profits for plywood are about five times higher than they are for raw logs and about 80 per cent higher still for furniture, an industry Indonesia wants to develop. By switching from logging to manufacturing plywood, one Indonesian company, Djajanti Group, created 10 000 jobs—four times the number of people it employed cutting logs for Japan.

Savings

When people put their savings in a bank account instead of buying jewelry or stuffing it under a mattress, something magical happens. Their money is lent to entrepreneurs, farmers, traders, industrialists, municipalities: the people who make an economy tick.

"Savings and economic growth constitute a virtuous circle," says Lee Tsao Yuan, deputy director of Singapore's Institute of Policy Studies.

Perhaps more than any other statistic, Southeast Asia's ability to save and invest tells its story of progress. In 1992 Indonesia's gross domestic savings rate (which includes government financial surpluses) was 37 per cent of gross domestic product. Thailand's was 36 per cent, Malaysia's 34 and Singapore's 47. Other regions saved only half as much: South Asia 17 per cent, sub-Saharan Africa 14.

Malaysia and Singapore employ mandatory provident funds—such as privately run pension plans that workers must invest in—to generate savings. But efficient banking systems help, too. The World Bank found that over the past 20 years banks in East Asian countries paid depositors an average interest rate 1.59 per cent higher than the rate of inflation. In sub-Saharan Africa, banks paid interest rates that were, on average, 11.13 per cent lower than the rate of inflation; in other words, they punished customers for putting their money in the bank.

Foreign Investment

The appreciation in the value of the Japanese yen in the mid-1980s, and again in 1994, has brought tidal waves of foreign investment to Southeast Asia, and with it technology, skills and market access. In 1991 alone, Japanese firms invested about $12-billion in Indonesia, Singapore, Malaysia, Thailand and the Philippines—double what they put into the United States.

But Southeast Asia's success story with foreign investments is not just about Japan. Between 1989 and 1992, Malaysia attracted $16.8-billion of foreign investment, most of it from outside Japan. One big investor, Thomson Consumer Electronics, now employs more people in Malaysia than it does in its home country, France.

Luck

For 25 years, Southeast Asia was able to attract labour-intensive industries with little competition. Through the 1970s and 1980s, China was uncertain about free market economics, Vietnam was sealed off from the world, India was following the protectionist path and Latin America was drowning in inflation.

This lack of competition coincided with two extraordinary economic events: one of the great export booms in history, and a period of almost unprecedented innovation in consumer products. The arrival of completely new product lines—hand-held video cameras, video-cassette recorders, laptop computers, boom boxes, compact disc players— created entire subindustries in search of a low-cost home.

The Future

For most of Southeast Asia, the easy part is over. The consumer boom has dissipated and there are plenty of new kids—Mexico, China, Vietnam, India—on the exporting block to fight for what's left.

"I am not that optimistic," says Mari Pangestu, an economist at the Centre for Strategic and International Studies in Jakarta. "We can muddle through another five years but beyond that we may have trouble with China and India."

Few would deny that Southeast Asia's next economic miracle will be much tougher to achieve. Its countries need to invest seriously in skills training if they want to attract higher-value industries. Although wages for both men and women are rising more rapidly than ever, gender gaps also remain a serious problem in such places as Bangkok, where female factory workers in 1989 earned, on average, 40 per cent less than male factory workers. And if the region doesn't reverse many poli-cies affecting the environment, one of the region's great assets— an abundance of natural resources—may soon be gone.

But with growth each year of 6, 7 and 8 per cent, the region's enthusiasm and expectations swell.

To cope with its low skills base—four million of its six million factory workers have only a primary education—Thailand recently increased its education budget by 13 per cent and allocated one-billion baht ($55-million) to training in science and technology.

As for China, Vietnam and India, Malaysia sees them as emerging markets, not emerging threats. One sign that it's right: China last year became Malaysia's No. 1 supplier of tourists. And many went home with computers, video cameras and CD players.

At Malaysia's state-owned automaker, Perusahaan Otomobil Nasional Berhad (Proton), managing director Mohamad Nadzmi Salleh exudes nothing but confidence. With the world's fastest-growing market all around him— Southeast Asia bought one million cars last year—he has set a new goal for Proton: "To one day replace Japan as the world leader in the production of the popular range of automobiles."

His target is not just a golden hope. Mr. Nadzmi recently told his Malaysian suppliers to cut their costs by 30 per cent. Most have. "It's amazing what the human mind can do," he says, "when it's pushed."

Malaria Story an Inspiration

~

BY

DAVID

SUZUKI

On the early 1960s, when malaria was aggressively pursued with treatment and mosquito control, 10 per cent of the world's population was at risk. However, with resistance of mosquitoes to insecticides and malarial parasites to drugs, the disease is now a threat to almost 50 per cent. Each year there are as many as 300 million cases, two-thirds of them new infections.

The statistics are grim. But an inspiring story about this modern plague has emerged from Latin America, which has 3 million cases.

In 1970, Brazil began construction of the Trans-Amazon Highway, thereby opening the forest to poor farmers, ranchers and gold miners. Malarial cases erupted more than 20-fold and 98 per cent of them are in the Amazon. Far more Brazilians die of malaria each year than from AIDS and cholera combined.

The life cycle of malaria offers hints to where it might be attacked. When an Anopheles mosquito sucks blood from a person infected with malaria, the parasites are stimulated in the insect's gut to mate and lay eggs in the gut wall. From the eggs, larvae or "sporozoites" hatch and are injected back into a person when the mosquito has another blood meal.

Once in the bloodstream, the sporozoites head for the liver, where they invade cells and multiply while changing into a new form called a "merozoite." The liver cells then break open and release merozoites, which then infect blood cells, multiply further, kill cells and infect still more.

Traditional attempts to develop vaccines used proteins of the infecting sporozoites to stimulate antibodies but never succeeded.

These failures prompted

Colombian chemist Manuel Patarroyo to take a radically different approach, focusing on the merozoite stage. Rather than purifying parts of the malarial parasite to induce immunity, Patarroyo chemically synthesized pieces of parasite protein he suspected were used to hook onto and penetrate cells.

And then instead of injecting different pieces one at a time, he attached four together (three from the merozoite and one from the sporozoite) and tested their combined ability to stimulate antibodies to stop malarial invaders.

And it worked! The world's first completely chemically synthesized vaccine confers immunity to malarial infection.

Patarroyo's success is a tribute to his brilliance, originality and diligence. The fact that he achieved it in a developing country is all the more extraordinary.

Yet, being in a scientific backwater may also have been a blessing. Patarroyo could afford to take bold risks. Part of his animal research was carried out in the Amazon at one-twentieth the cost for comparable work in the U.S. Using skilled technicians on Colombian wages, he could synthesize large volumes of pure proteins hundreds of times more cheaply than in a rich country.

The cost of the three shots of Patarroyo's vaccine per individual is about 30 cents.

Patarroyo encountered initial skepticism and criticism. But once his experimental test procedures and chemical purity were assured and the vaccine was still shown to be effective, the accolades rolled in.

He told me one major multinational drug company offered him $68 million and a commission on all sales for the rights to his patent. Patarroyo turned it down because he doesn't believe profit should come from suffering and poverty.

With total support of his family, research colleagues and the president of Colombia, Patarroyo gave the rights to the patent on his vaccine to the World Health Organization. He imposed a stipulation: the vaccine must be manufactured in a developing country and if Colombia can do it cheapest, the facilities will be located there.

This astonishing and rare humanitarian act is an inspiration for all people. It is fitting that Manuel Patarroyo has just received the prestigious Prince of Asturias Prize from Spain of $38 000 (U.S.) and an original Miro painting.

All of humankind salutes him.

Children Are Color-Blind

BY GENNY LIM

I never painted myself yellow
the way I colored the sun when I was five.
The way I colored whitefolks with the "flesh" crayola.
Yellow pages adults thumbed through for restaurants,
taxis, airlines, plumbers...
The color of summer squash, corn, eggyolk, innocence
    and tapioca.

My children knew before they were taught.
They envisioned rainbows emblazoned over alleyways;
Clouds floating over hilltops like a freedom shroud.
With hands clasped, time dragged them along and they
    followed.

Wind-flushed cheeks persimmon,
eyes dilated like dark pearls staring out the backseat
    windows,
they speed through childhood like greyhounds
into the knot of night, hills fanning out,
an ocean ending at an underpass,
a horizon blunted by lorries, skyscrapers,
vision blurring at the brink of poverty.

Dani, my three-year-old, recites the alphabet from
billboards flashing by like pages of a cartoon flipbook,
where above, carpetbaggers patrol the freeways like
Olympic gods hustling their hi-tech neon gospel,
looking down from the fast lane,
dropping Kool dreams, booze dreams, fancy car dreams,
fast foods dreams, sex dreams and no-tomorrow dreams
like eight balls into your easy psychic pocket.

"Only girls with black hair, black eyes can join!"
My eight-year-old was chided at school for excluding a blonde
from her circle. "Only girls with black hair, black eyes
can join!" taunted the little Asian girls, black hair,
black eyes flashing, mirroring, mimicking what they heard
as the message of the medium, the message of the world-at-large:
        "Apartheid, segregation, self-determination!
        Segregation, apartheid, revolution!"
Like a contrapuntal hymn, like a curse that remains in
a melody trapped.

Sometimes at night I touch the children when they're sleeping
and the coolness of my fingers sends shivers through them that
is a foreshadowing, a guilt imparted.

Dani doesn't paint herself yellow
the way I colored the sun.
The way she dances in its light as I watch from the shadow.
No, she says green is her favorite color.
"It's the color of life!"

Revolution!

BY

VIC

FINKELSTEIN

t all began when Shanti was late again for school. The rain was to blame. It had churned up the mud so much that she had had to drag her lame leg way beyond the usual river crossing to find a safer spot. Crossing the river was harder than she'd expected. She reached school, exhausted, muddy, wet—and pleased with her success. But the teacher, Mr. Pahad, had shouted at her. "You silly child! You should stay at home when it rains."

He was in a bad mood today—not helped by the fact that he knew he would probably feel compelled to help Shanti get home after school. That would mean getting mud on his suit—and tonight he was going to a banquet to receive an award for his contribution to the welfare of the disabled children at his school.

A thought came to him. What if all disabled children, adults and older people, were sent to a special village where everything was designed to meet their special needs? There would be special schools, special sport, special employment and special housing.

Whenever he had a spare moment during the rest of that day Mr. Pahad re-wrote his speech for the banquet. Then he had another brainwave. He would take Shanti with him. She looked so sweet and shy with her muddy dress and twisted foot. He would explain her plight and ask for donations to build a disabled village for all the disabled people of the district.

So a dazed Shanti found herself at the banquet, being waited upon by servants in spotless uniforms. She felt embarrassed by the mud on her dress and wanted to use the washroom but Mr. Pahad insisted, "No, no. I want everyone to see you the way you are." Mr. Pahad made his speech and, indicating Shanti, explained his proposal for the disabled village. People liked the idea. They clapped and cheered. Later they arranged committees for fundraising. It was decided that the village would be built on the other side of the river and that it would be a village for wheelchair users. If it worked other villages would be built for the blind, deaf and people with learning difficulties.

By the end of the evening Shanti had been generously patted and kissed but never actually spoken to. In the weeks that followed her picture appeared on a large fundraising poster all around the district. The poster read "Shanti and others like her need your help. Give generously for the disabled village."

On the day the village was opened flags flew, there was food for all. The disabled people were lined up to meet famous guests. Speeches were made and more awards given to Mr. Pahad. When the festivities were over the able-bodied caregivers and professionals established a convenient (for them) routine to get the residents out of bed in the morning and back into bed in the evening. During the day the villagers had occupational therapy.

At first the villagers hated being taken away from their homes. But as time passed they found that being together had certain advantages: they could meet more easily and share ideas and feelings. Shanti liked being in a place where everything was arranged for people who lived in wheelchairs. You could do all sorts of things which you were prevented from doing "beyond the river" as the able-bodied world was now called. You could even do your own shopping. This made the residents think: if they could do their own shopping why should they not work in the shops themselves too? They suggested this to Mr. Pahad. But he said it was "unrealistic." So were most of the other suggestions residents made. Relations between staff and residents got worse and worse as one suggestion after another was trashed by the able-bodied authorities.

The villagers got together to discuss the situation. They debated for several hours. Some felt that with the force of argument they could change the attitudes of Mr. Pahad and his helpers. Others felt there was no point—that able-bodied people didn't begin to understand the experience of disability. The villagers finally formed a committee and began plotting revolution.

A few days later, when most of the helpers had gone over the river for their regular monthly staff meeting with Mr. Pahad, the disabled residents took direct action. They barred the village gates, closed off all entrances and exits and flooded pathways leading to the village.

Shanti, being the smallest and lightest, was lifted onto a wall where she could watch for the returning helpers while disabled villagers prepared to do battle. "They're coming, they're coming," Shanti screamed with excitement. From where she sat she could see everything.

First the helpers were surprised. They never expected disabled people to do anything without their assistance. Then they got angry as their feet got wet and muddy in the pathways and they found all the entrances closed. When they realized that their jobs were on the line the helpers became all the more convinced that "the disabled" urgently needed their help. They broke down the gate and rushed in—only to bash their heads and fall flat on their backs. A low ceiling of poles had been tied into place with just enough room for wheelchair users to move freely underneath but too low for the "walkers"!

Then a row of villagers moved forward, pushing the dazed caregivers out of the village with scoops that had been fitted to the front of each wheelchair. Eventually the helpers gave up and left the village.

There were many changes after the Revolution. Roads and paths were dug up and replaced with wheelways. Doorways and ceilings were lowered to a more reasonable height for wheelchair users. The shops, the school, and places of employment were all altered. Fashion became more interesting as the village shoe shop began selling multicolour designer tires for wheelchairs.

As Shanti grew up the memories of the able-bodied soon faded and the villagers forgot that they were supposed to be disabled. In this village they were the "normal." Life went on peacefully for several years until one day the villagers were once again brought face to face with the able-bodied from across the river.

It happened during a particularly heavy rainy season when the river burst its banks and flooded the able-bodied village. Those who escaped made their way to the nearest high ground—which just happened to be the disabled village.

Shanti was busy making a pot when she spotted the first flood survivor on the main wheelway. Then another, and another. A whole stream of able-bodied people poured into the village, getting their feet stuck in the wheelway tracks, and knocking themselves out on the doorways as they stumbled into houses, looking for shelter. Soon the village doctors had their hands full. Other villagers prepared food for the victims—who had to eat off the floor because the disabled villagers had long dispensed with tables. Their own wheelchair attachments were suited to all uses.

The disabled villagers felt sorry for the able-bodied. They seemed so clumsy, helpless. Many of them couldn't even get out of the old community centre building that had become their residential home without damaging their feet in the wheelway tracks. Special transport was devised so that a little trolley for the able-bodied could be attached to a wheelchair.

But who was going to look after the able-bodied? They could not work as everything was designed for people in wheelchairs. Soon able-bodied people in ill-fitting clothing made for wheelchair users were to be found on wheelway corners begging for food and money. More fortunate able-bodied refugees were taken from their residential home to the day centre where they could do some basket work and other useful occupational therapy.

The biggest problem for the medical profession was the chronic bruising of heads. The village doctors diagnosed this as "cerebral indigene" and recommended either a harness to keep the able-bodied bent double at wheelchair height or padded guards which were strapped to the forehead.

Shanti was becoming increasingly concerned about the welfare of the able-bodied. She was asked to organize a public appeal for money to provide the able-bodied with "care in the community." Then someone suggested that instead of care the money might be used to set up a special place where the able-bodied could live. This reminded Shanti of Mr. Pahad's original scheme for the disabled village—and she was opposed to it.

"Disability," she protested as she addressed a public meeting on the subject, "isn't something that you have. It is something that happens when one group of people create barriers by designing the world only for their style of living."

And she went on: "We will not make any progress by keeping disabled people on one side of the river and non-disabled people on the other, with each side creating barriers. What we need is to build up the banks so that the river does not flood and to build bridges across the river so that we can meet, exchange experiences and create an environment where we can celebrate human difference." She had become quite an idealist, had Shanti.

The Cardboard Room

BY

TERESA
PITMAN

M y parents hate a lot of people.

It's not like kids, not like your best friend when you were six, saying "I hate you" because you won't give her a ride on your new bike, and stomping away home and then tomorrow, you're best friends again. Kids can be mad at each other and then it's all over an hour later.

Adults aren't like that. They hate deeper, stronger, and they cover it all over with smiles and soft words.

Oh, my parents have their reasons. They don't like the Johnsons down the road because they're Americans and if they really cared about this country, they'd get their citizenship. And because Americans are trying to take over the country, not by fighting a war, but through secretly buying up all our property. The Johnsons' split-level is just one small step in somebody's master plan, I guess.

And the Freedmans might be okay if they didn't drink so much. And the Archers have six kids when, as my mom says, "it's totally irresponsible to have more than two children in today's world."

It doesn't matter if the people try to be nice to them. I remember going to a meeting at the school, all of us together, and Phil Martin's dad came over. We laughed at Phil, you know, because of his clothes, and now here was his father, dressed no better.

He had this smile on his face that looked like it hurt him to keep it there and the skin on his forehead was streaked with sweat. When he asked my dad if there might be any jobs up at the plant, my father just smiled and said no, but he'd keep his eyes open.

Even before Mr. Martin had finished saying thank you, my parents had started to walk away and all the way home, they shook their heads over the man's nerve, asking right out like that.

They didn't hate me then. None of this would have mattered—everything would still be okay—if it hadn't been for Eric.

"That's the problem with the public school system," my dad said after that first Saturday at Eric's house, and my mother scowled at him.

"I told you we should have sent her to a private school," she said and then shrugged her shoulders in resignation. It was an old argument and nobody wanted to hear it again. She turned to me instead.

"What was their house like? Did you eat any of their food?"

I was afraid to admit I'd eaten anything. It was only an apple, but I'd probably have had to stay in bed the rest of the day with a thermometer in my mouth if my mother knew. And what could I say about the house?

"It's small."

All the way home—I had to walk, Eric's family only had one car and his father worked on Saturdays, too—I thought about their house, thought about how my parents would laugh if I told them about it, how it would become a joke to share with Dad's friends from work when they came over. And all I could say was, "It's small."

It *was* a small house for them, the parents and the four kids, and it wasn't until I was inside that I discovered Eric's grandparents lived there, too. And even though it was small, the rooms looked stark, half-empty: They had almost no furniture.

The kitchen had no space for eating and there wasn't a dining room at all, so their big table with eight mismatched chairs around it stood in the middle of the living room. I guess they figured they might as well put it there, because they didn't have a sofa, only a faded armchair and a black-and-white TV set.

My parents would have laughed at that. And they would have

laughed at the bedsheets hung in the bedroom windows and the jars of beans and pickles and home-canned fruit on the basement shelves.

"Hasn't she ever heard of grocery stores?" my mother would have said. "They'll probably all die of food poisoning." And they would have laughed at Eric's room.

It started out, you know, as a physics project. Mr. Hennessy announced that he was assigning partners for this one because he was tired of the same people always working together. I don't see what difference it made to him: We like it that way.

For a few minutes I had dreams of doing the project with Mike McCann who sat in the back and was on the football team and always kept a bottle in his gym bag in case the class got boring. But Mr. Hennessy assigned me to Eric Nye.

"I don't see what you're complaining about," he said when I asked him, after the bell rang, if I could switch.

"Eric's probably the best student in the class."

"I'd rather work with another girl." That wasn't completely true, but it would be better than Eric.

"Get out of here. You do the assignment with Eric or you get zero."

And my dad said, when I told him, that it was teachers like Mr. Hennessy who were the problem with the public school system. They had no choice but to let me go, the next Saturday, when Eric asked me over to his house. He had some magazines and stuff, Eric said, that we could use for the project.

Maybe it was mostly curiosity that made me go. The Nyes were— well, even the people my parents couldn't stand avoided the Nyes. There was a war in their country and they had come here with no money and no job. Refugees, some people called them.

My father had other names for them. They had come here, he said, to be supported by our welfare system—OHIP and everything. When he found out that Eric's father was working two jobs, weekdays and weekends, he was even madder: "No wonder we've got so many Canadians unemployed."

Once you get used to hating one person, one family, one group of people, somehow it seems to get easier and easier to expand that group, to include more and more people in it. Practice makes perfect or something like that.

When I knocked on the door, Eric let me in and introduced me quickly—I didn't remember any of the names—and took me down to the basement. The house had three bedrooms, just like ours, but his grandparents had one and his parents another and so that Eric could have a little privacy, a place to work, they gave him a corner of the basement.

It was just a regular, unfinished basement with a cold floor and cement wall and a bare lightbulb in the middle with a string tied to the chain. And one corner was enclosed with cardboard. That was Eric's room.

His father had taken apart some of those big cardboard cartons that stoves and dishwashers come in (they must have picked them up from the back of a furniture store because the fridge and stove they had upstairs hadn't been new for years), and stapled them together to make walls.

It was like the houses little kids make out of shoeboxes, with a cardboard door cut on the side nearest the stairs. And when we went inside, all Eric had was a mattress on the floor and some boxes with clothes and books and papers in them. I couldn't see anywhere to sit.

Eric started rummaging through a box full of papers, but stopped when he noticed me standing there awkwardly.

"Just sit on the bed."

I pulled up the wrinkled sheet and smoothed out the blanket. If this had been me with Mike McCann, sitting side-by-side on his unmade bed, I might have been self-conscious, wondering about the possibilities.

But Eric wasn't the kind of person you could have fantasies about. He was so thin, for one thing. I could see the line of his collarbone poking through the worn material of his shirt, and the faint shadows of his ribs. But part of me wondered what it would be like to touch him, his golden-brown skin, wondered if it would feel different, wondered if it would be warm or cool, smooth or dry. Not fantasies. Just wondering.

"Here," he said finally, handing me a bundle of papers. "I photocopied these from some magazines at the library."

Before I could look at the articles, he took them back and read them out loud to me. It wasn't that the words he was reading didn't make sense. It was great material—we were going to have the best project in the class. It was the way he was reading it, the tone of his voice. This was physics, after all, school work.

"You really like this stuff, don't you?" I asked him and when he stared at me I knew he couldn't even understand why I had asked.

After I'd been there three or four times on Saturdays, putting this project together, his mother invited me to stay for lunch (I told my parents I was going to Burger King with Sylvia). I ate the food without looking too hard, just swallowed everything quickly before I could taste it too much, and tried to smile at the grandparents who didn't speak English.

I was getting used to Eric's mother who was so quiet and seemed to spend almost all her time in the kitchen, and after lunch, I offered to help with the dishes. And while we were sloshing around in the soapy water, I found myself asking her the things I was afraid to ask Eric.

"What was it like to leave your own country and come here?"

Silence. I shouldn't have asked.

"Like being born."

Silence again. I clinked the glasses together under the water.

"The pains, you know, come stronger and stronger, closer and closer, until it becomes too much to bear. And finally you have no choice but to be born into this new country. The cord is cut. This is our home now."

She wasn't looking at me. I wanted to ask her if she knew that people here hated her, hated her family, but when I saw her face, I didn't need to ask. She knew. They all knew.

That was the beginning. And maybe everything would have been all right had I let it stop there. But I didn't.

We finished the project and handed it in. I just didn't mention it to my parents. And on Saturday morning, I knocked again at Eric's door. He didn't ask me what I was doing there, just let me in. And we sat on his bed in that cardboard room and talked.

He didn't talk about the things Sylvia and the other kids did. He didn't care about clothes or music or long stories about getting stoned and driving around the city at twice the speed limit.

He talked about all the things he wondered about, like what colors looked like to different people, and whether God really existed, and about the things he was afraid of. And sometimes he would tell me stories about the country they had left behind, and about his two brothers who had died, the older one who was killed in the fighting and the baby who was sick when there was no room in the hospital.

I shook my head, not disbelieving him, but unable to make his stories fit into the world that I lived in, where there was always room for babies in the hospital and older brothers went off to college, not to war.

"You don't see how fragile it all is, do you?" he said. "You don't know how quickly everything can change and then nothing is ever the same again. Never the same again."

I went home that day thinking I'd never go back, but I did. It was like having a tooth with a cavity that you are trying to ignore, but your tongue keeps going back to that tooth and probing and pushing at it. Talking to Eric was like that. I needed to explore the exact dimensions of the cavity, find the depths of the pain.

But I thought I could do it without anyone knowing. It worked for a while.

"You've been spending a lot of time with that Nye boy, haven't you?"

"Well, there's that physics project...."

"That's not what I mean."

I felt my face flush. They must have found out about the day we skipped school together. Eric wanted to show me a new exhibit at the museum, but we left by separate doors and met two blocks away from the school so that nobody would know. I hadn't told anyone. They must have found out.

But my mother flattened the paper she held on the table in front of me. Physics project. Graded A+, with "Excellent work" added in Mr. Hennessy's red ink.

"You finished this weeks ago." Calmly, coldly.

"Eric..."

"What have you been doing with that boy all this time? Are you sleeping with him? How could you do this to us?"

I knew it didn't matter what I said. Suddenly I knew what Eric meant about things changing. The rest of the conversation, the accusations and the threats were not important: It was just that everything had changed.

Now I could see the walls. It was as though I had taken a piece of photographic paper, completely blank, and dipped it into the developer and watched the picture spring into focus. Now I could see the walls my parents had built: solid brick and mortar, impenetrable. And now I was on the outside.

My parents hate a lot of people. Oh, they have their reasons. It doesn't matter if you try to be nice to them. It doesn't matter...

Eric knew, Eric knew all along how fragile things are, how easily destroyed. And nothing is ever the same again.

Children behind fence that separates them from white community near Johannesburg, South Africa, 1973

from

Daughters
of the
Twilight

BY

**FARIDA
KARODIA**

T he weeks dragged by. Things were quiet and I was bored. Weekdays weren't too bad, but I was beginning to get fed up with school too. Weekends were even worse, because there was nothing to do. I had read every book in the house, including the dozen or so fly-encrusted romances that were strung up in the front window and had been there ever since I could remember.

One Saturday morning, for want of something to do, I volunteered to pick up the post at the post office.

I returned with the pile of letters. I had already checked and there was nothing for me. I wasn't expecting a letter, yet I always went through the same exercise, sorting through the letters in the hope that there would be one for me. But it was always the same. Nothing. Yasmin's letters were addressed to Ma. Anything for me was usually enclosed in her letter.

Ma riffled though the letters and picked up a large manilla envelope.

She frowned and carefully scrutinised the government's identification stamp as though it would provide some clue to the contents.

There was a long silence while she read the letter through carefully. I noticed the sudden rush of blood from her face and knew instinctively that something was wrong. I waited for her to say something. Papa noticed too.

"What have you got there?" he asked.

"It's a letter from the Group Areas Board."

There was a moment of silence. Ma looked up.

"What do they want?"

"They've assessed our property," she said distractedly, scanning the typewritten page again."

"What? When?"Papa asked, startled.

"1,200 rands...According to this, the house and shop are worth 1,200 rands."

"Are they mad? When did they assess it?" Papa demanded.

"They don't say. But I think it was the time that Afrikaner came by to inspect the property about four or five months age."

"Why don't I know about it?" he asked.

"You were the one who spoke to him. Remember, you thought he was here to assess for property taxes?" she reminded him.

Papa had forgotten all about the incident. He nodded as it came back to him, his eyes narrowing. He had asked the man what he wanted but had got no answer from him. At that time they had thought nothing of the episode, but now it took on new significance.

I listened as my parents rehashed the incident, reading all sorts of relevance into how the man had walked, looked around and refused to answer questions.

"They're going to take over our home and our shop." Ma was aghast. She glanced from Nana to Papa in stunned silence.

"Not while I have a breath left in my body!" Papa declared. Then he laughed. "This is just another bureaucratic bungle. It's a mistake. We won't worry about it."

"What if it isn't?" Ma asked.

He sat down heavily, considering her question.

"What are we going to do?" she repeated, thinking that he hadn't heard her.

"I don't know! But I won't let them take what we've worked for all these years! This is our house. We've built this business from scratch.

There was nothing here when we arrived. We'll put up a fight. You just wait and see." Papa's eyes darkened. Along his temple a vein stood out like a knotted rope. His clenched hand jerked open and involuntarily twitched as it rested on the top of the old rolltop desk.

"Fight with what?" Nana demanded. "They'll come with bulldozers and flatten all of this whether you're in it or not. I've seen how the Group Areas Board operates. They declare an area white, then they come in and take over. They're not interested in owning these buildings. It's you they want out of this area. They don't care about your life or what you've put into this place. They don't care about anything except getting you out."

"I don't think they'll break this down. It's a solid structure. I think someone wants it—probably old Faurie or van Wyk," Papa said.

"We should have seen it coming, especially after what they did in the other small towns," Ma said. "Still, I think Abdul is right, we should put up a fight. We can't just go like lambs to the slaughter."

Tormented, Papa leaned forward in his chair, drawing a hand over the bald patch on his head. "What else do they say?"

Ma exchanged troubled glances with Nana, "Nothing more, except of course that the property is worth 1,200 rands…"

Papa came upright so quickly that the old spring in the swivel chair twanged. "It's worth a lot more than 1,200 rands. Look what we've done to the place. Look at all the improvements!"

"We'll discuss this when you've cooled off," Ma said.

"Why steal our property. They can just ask us to give it to them for nothing," he said with a touch of sarcasm.

"In the end they'll do just that," Nana remarked.

Her words incensed Papa even more.

"Mum, please," Ma muttered.

"This place is worth a lot more than 1,200 rands," he continued. "I can tell you. We built all of this from nothing. We sank all our money into the business and this property." There was an angry pause. "Go on, what else do they say?" he demanded.

"They've given us six months to find another place on our own, or…the alternative they present here is to move to McBain, which is a little less than halfway to Queenstown."

"I know where it is," he snapped. "It's in the bush. A pile of bricks in the veld beside the road."

"We've passed by it hundreds of times," Nana said, "never giving it a second glance. There's nothing there. Abdul's right. It's just a pile of bricks."

"Dear Lord," Ma sighed wearily. "My home...Both my children were born here. I love this place." She drew her hand cross her face.

"We're not moving. This is our home and we're staying right here," Papa told them.

Something really frightful had happened to the family. I stood to one side watching my parents and I wondered how such a terrible dread could ever be dispelled. The fear and anxiety of a future filled with uncertainty was unbearable.

Ma nodded. "We'll see a lawyer. Abdul's right. We'll fight them. Why should we give up our home? Our livelihood is tied up here."

They were at the lawyer's office first thing when it opened on Monday morning.

He told them that there was nothing they could do. It was a law, an act of parliament, that each racial group be confined to its own area. He said that we had no alternative but to abide by any decision the Group Areas Board made.

They came home angry and disappointed.

"That man is a mangpara," Papa said as they stepped in the door.

Nana and I knew instantly that things had not gone right.

The adults talked of nothing else but this new threat.

"They leave one with nothing," Nana sighed, "not even your dignity."

"What's happening?" I asked.

Ma shook her head wearily. They were too preoccupied to explain it all to me.

"What's going to happen?" I persisted.

"Everything will be taken care of," Papa answered.

My parents and grandmother latched on to the phrase, taking refuge behind it whenever they became impatient with my questions. I wished that Yasmin was here. I missed her. I had no one to talk to now.

～

The Men from the Group Areas Board came on the day that Papa and Daniel had gone to East London. From the front door Ma and I watched them pulling up in the police van, accompanied by the sergeant of police, two constables and their dogs.

"What is it?" Ma asked. At first we thought that they had brought bad news about Papa. But when Ma saw the guns and the dogs, the blood drained from her face.

"What do they want, Ma?" I asked.

My mother shook her head. She didn't know either.

"These men are from the Group Areas Board," Sergeant Klein told us, gesturing towards the dark-suited men. "They...we are to enforce your eviction."

Ma's face was stark with fear. Then she looked at Sergeant Klein. It was all a mistake, she decided, expecting that the sergeant would rectify the ghastly error.

I thought so too. After all, he knew us, knew that we were harmless. Ma laughed mirthlessly, but there was no response from the circle of cold and dispassionate faces.

Nana, who had heard the commotion, came to investigate. "What's going on here? Has something happened to Abdul?"

Ma shook her head.

"What's going on, Meena?"

"I don't know, Nana."

Nana's questioning glance flew to the dark-suited strangers and then to the constables, finally coming to rest on Sergeant Klein, who was our only salvation. But he was studying the tips of his boots.

"We're locking up this property," one of the men said.

Nana was aghast. "Now?" she asked.

The man nodded.

"You can't do that. You're supposed to give us notice. We know the law. Besides, her husband isn't here. He's in East London," Nana started to explain.

I shook my head.

"You received your eviction notice some time ago," Sergeant Klein replied.

I exchanged troubled glances with my grandmother.

"No we didn't," Nana put in.

"We have a copy of that letter."

"But we didn't receive it," I cried.

"The letter was registered and we have documented proof that you got it."

I looked at Ma and it all became quite clear. Papa. It was the only explanation. He probably received the notice and destroyed it. Nana was right. He had been behaving very strangely.

"You've had enough time. Now move out of the way," one of the other men said.

Throughout this exchange Sergeant Klein stood to one side, staring at the wall behind us.

"You've left us no choice," the first man said. "The matter is now in the hands of the police."

The drumming of voices and the clattering in my head made it almost impossible to hear what was being said. This is Sterkstroom, I thought. This can't possibly be happening here. This is not a big city. People here are not evicted.

"We're not ready. We need time."

"Go phone your Papa," Ma instructed. Her bun had come undone; her large anguished eyes were turned on Sergeant Klein, pleading. Then her hands dropped to her sides in a gesture of helplessness. This was the image of my mother I took with me as I hurried away to phone Papa.

When I returned, two of the younger constables had pushed their way into the house.

"Get out! Get out!" Nana pressed her hand to her chest. "Leave us alone!" Then slowly, her back supported against the wall, she slid to the floor. She was breathing heavily, her face ashen. Ma and I helped her to the door, where one of the younger men roughly pushed us outside.

"Pas op!" the sergeant cried, startled as Nana staggered. She would have fallen had Mrs. Ollie not put her arm out to steady her.

In the passageway one of the young constables was dragging the blue stuffed chair from the bedroom.

"What are you doing?" Ma asked, horrified as he tossed it on to the sidewalk.

Nana covered her face with one hand, the other hanging limply at her side. Ma put her arms about Nana, supporting her while I righted the chair.

"Heinie, heinie, Khoskhaz!" Gladys shouted from the kitchen.

Ma helped Nana into the chair.

"Did you speak to your Papa?"

I nodded.

"Well, what did he say?"

"He said not to do anything. He's on his way."

"Fine thing after the mess he's made." Ma muttered. "Keep an eye on your Nana while I go and see what's happening to Gladys."

Nana's face was expressionless, one side pulling downwards. I sensed that something was wrong, but I didn't know then that she had suffered a slight stroke. I squeezed her hand reassuringly, dabbing at the spittle which dribbled from the corner of her mouth. The police would soon be gone and then we'd be able to move back in again.

"No! No! Please!" Ma's cry startled me out of these reflections.

I rushed to the kitchen. My mother was clinging to the arm of one of the constables, who was struggling to free himself. Ma held on as though her life depended on keeping him at bay.

Lying on the floor were pieces of broken porcelain.

"They belonged to my grandmother. Please leave them alone. Leave me alone! I'll do the packing. Why are you doing this to us Sergeant Klein? Why? This is our home!" she cried.

"I'm sorry Mrs. Mohammed, I'm only doing my job," he muttered, and walked away.

But Mrs. Ollie stopped him. "What are you doing to these people?" she demanded. "I know them. You know them too. They're not criminals. Why are you treating them like this? In God's name, man, what are you doing?"

"Look," Ma cried, spreading her arms. "My tea service, look at it," she said, choking on a sob. Suddenly there was a loud crash from the bedroom. We all rushed to the front. On the floor were the fragments of Ma's precious porcelain basin and jug.

"Oh God, no!" Ma picked up the larger pieces, holding the shard with its delicate pattern of blue forget-me-nots against her cheek. The other constable placed his hand on her shoulder, apparently intending to guide her out of the room, but she jerked free.

"I'll get her out," Mrs. Ollie said anxiously.

Sergeant Klein nodded.

Ma leaned against the dresser, clutching the piece of porcelain, tears streaming down her cheeks. Mrs. Ollie led her outside. "This belonged to my grandmother," Ma whispered bleakly.

"Kom nou, Delia. Kom," Mrs. Ollie whispered, glaring at the police.

Out on the sidewalk Ma stared vacantly at our scattered effects.

Many of the townspeople had gathered. Some of them helped, others stood around, uneasy witnesses shuffling from one foot to the other.

"Are you all right, Mum?" Ma asked.

Nana nodded with great difficulty.

"I don't know what's wrong with her," I said.

"It's the shock."

"Come over to my place for a moment," Mrs. Ollie urged Ma and Nana.

Nana's head teetered. Ma frowned, her troubled glance studying Nana. I wrung my hands. I had seen that look in Nana's eyes, an expression of unspeakable terror.

"Take care of your grandmother," Mrs. Ollie instructed. "I want to take your Ma inside for a moment. I want to get her a cup of tea. It'll help you to pull yourself together so you can think about what you're going to do."

"I don't like the way my mother looks. I should get her to a doctor," Ma said.

"Dr. Uys is out of town. I'll get her a cup of tea."

Some of the bystanders helped to pick up our scattered clothing, which I hurriedly threw into cardboard boxes.

Mrs. Ollie showed Ma indoors to a chair by the window.

When the tea was ready she stuck her head out of the front door. "Kom Meena, vat vir you ouma n'lekker koppie roibos tee."

I dropped what I was doing and went to fetch the cup of tea for Nana.

"Dankie," Nana said, accepting gratefully. Her mouth was not too bad now. She wasn't dribbling any more and I noticed that she was able to move her arm a little.

I held the cup to Nana's lips. Mrs. Ollie had served the tea not in the enamel mugs they used every day but in her best china.

I returned the cup. Mrs. Ollie and Ma were sipping their tea in silence, Ma with her head bowed, supporting the cup and saucer in her lap.

"Under the circumstances you might find this hard to believe," Mrs. Ollie said in Afrikaans, "but we're not all like that." She gestured to the police. "I've been your friend for a long time. I know what you're going through. Hardship and pain are the same whether you're white, brown, black or green."

"You've been a good friend to me all these years Sinnah," Ma smiled sadly. "Sinnah Olivier...I'd almost forgotten your last name. All these years you've been Ollie because the children couldn't say Olivier."

I was about to leave when Ma said, "Keep an eye on Nana please, Meena."

Sounds from outside carried indoors and soon her eyes welled up again. They were emptying the store in the same manner as they had emptied the house.

"You can store your stuff in my shed. The cows will be all right outside," she said, accompanying Ma to the stoep.

"Thank you," Ma said. "Thank you so much. I'll leave the big items here...for a short while anyway."

"What are you going to do, Delia?"

"I suppose we'll have to move to McBain. There's nowhere else for us to go."

"That's ridiculous. The place is nothing but ruins."

Ma shrugged.

Mrs. Ollie sighed. She studied Ma. She didn't have to say anything. It was all there in her eyes.

Ma turned to go. "Thanks for the tea...and everything," she said, offering her hand, but the Afrikaner woman ignored her outstretched hand and embraced her, right there in the middle of the street with half the town looking on.

"Good luck," she said.

"There's a phone call from East London," someone called from the doorway.

"It must be Abdul. I'd better go."

I rushed after her. Ma paused in the doorway to the store. The police were too busy carrying out their nefarious deeds to notice us. Ma took the phone and covered her ear to shut out the commotion. She watched them with a disaffected air, as though she had cut herself off from all that was happening here. In stunned silence she listened to the voice on the other end. One hand flailed behind her as she groped for a chair.

"What is it, Ma?" I asked.

"It's Aishabhen. Your Papa's had a heart attack," she said, motioning for me to come closer.

I took the phone from her.

"The doctor says he'll be all right. The attack came on just after you called this morning. I think it was the shock of what was happening there," Aishabhen said.

"Tell her I'll take the train. I'll be there tomorrow morning," Ma said. Then, as an afterthought, she took the phone. "Let me talk to him."

"That won't be necessary," Aishabhen told her. "You have enough to deal with right now. He'll be fine."

"Are you sure, Aisha?"

"Yes, I'm sure. He won't be in hospital for long. He's looking fine, Delia. Don't worry. He should be discharged tomorrow. He can stay with us for another week or so while he recovers."

"I really appreciate this."

"Are you all right?" she asked.

"Yes. My mother wasn't too well, but I think she's feeling a little better now. We'll be leaving for McBain in the morning."

"Let me know if you need anything. Would you like Farouk to come and help you?"

"No, we'll be fine, thank you. You've been a great help," Ma said. Drawing a hand wearily through her hair, she put the phone down.

With the weight of Papa's illness off her mind she was able to think a little more clearly about what we would do once we got to McBain.

"Will he be all right?" I asked, eyes bright with anxiety.

"Aishabhen says he was lucky it was a mild attack. She doesn't think it's anything to worry about. We'll call the hospital later," Ma said. For a moment she watched the police carrying the merchandise out of the shop, then she sighed, shuddered and turned away. She walked away from the shop towards the house, changed direction and left the property through the side gate.

Most of our belongings and the stock from the shop were stored in Mrs. Ollie's garage. The items which we needed immediately were packed into the old Buick, repossessed from Mr. Erasmus.

Finally, Ma made arrangements for the transportation of the rest of our belongings to McBain by ox-wagon.

Mr. Petersen, principal of SAPS, took us in for the night and Ma was able to take Nana to the doctor. Dr. Uys's cursory examination revealed that Nana had had a mild stroke.

Gladys said that she would join us later when things were more settled. "What about Daniel's stuff?" she asked.

"We'll leave it with you. I don't think he wants to come to McBain," Ma said.

Early the next morning, before leaving the town, we stopped by the house. The doors were all padlocked.

Gladys stood on the front sidewalk until the car turned the corner at the end of the street.

We were leaving behind us not only our home but also a big chunk of our lives. Tears slid down Ma's cheeks as she watched Gladys's for-lorn figure in the rear-view mirror.

I turned around for a last look. Both she and Daniel had been such an integral part of our lives, one of the many threads woven into the fabric of our existence.

"There's no use dwelling on the past. We have to go on," Ma said, brushing the dampness from her cheeks.

"God, some day they'll pay for this," I muttered.

"Not them. We're the ones who pay," Nana said.

"What will happen to Daniel when he gets back from East London?" I asked.

"I don't know. I suppose he'll come to McBain...I don't know, Meena. I don't know anything, anymore."

No Longer
Our Own Country

~

BY

TANURE

OJAIDE

We have lost it,
the country we were born into.
We can now sing dirges
of that commonwealth of yesterday—
we live in a country
that is no longer our own.

Our sacred trees have been cut down
to make armchairs for the rich and titled;
our totem eagle, that bird of great heights,
has been shot at by thoughtless guardians.
Our borders have been broken loose
to surfeit the exotic appetite for freedom,
our flag ripped off by uncaring hands.
Counting the obscenities from every mouth,
the stupor, the deep wounds in our souls,
you can tell that we live in a country
that is no longer our own.

Where are the tall trees
that shielded us from the sun's spears,
where are they now that hot winds
blow parching sands
and bury us in dunes?
Where are those warriors
careful not to break taboos
who kept us from savage violations,
now that we face death?
Where are the healers
who offering themselves as ritual beasts
saved their neighbours from scourges?

We will expect in old age
to climb the mountain of prosperity
which we blew up in adolescence.
Our own country was a dream
so beautiful while it lasted,
and now we are exiles
in a country that was once ours—
we were born into another country,
a world that has gone
with a big boom.

But we will not perish in this other country.
We have lived through death
to this day,
we have deposed ourselves
and depend on alms that come our way.
We now know
what it is to lose our home,
what it is to lose a hospitable place
for this exile.
We expect the return of good days
and wiser, will no longer
let them pass from us.
For now we live in a country
that is no longer our own.

Walking Both Sides of an Invisible Border

BY ALOOTOOK IPELLIE

It is never easy
Walking with an invisible border
Separating my left and right foot

I feel like an illegitimate child
Forsaken by my parents
At least I can claim innocence
Since I did not ask to come
Into this world

Walking on both sides of this
Invisible border
Each and every day
And for the rest of my life
Is like having been
Sentenced to a torture chamber
Without having committed a crime

Understanding the history of humanity
I am not the least surprised
This is happening to me
A non-entity
During this population explosion
In a minuscule world

I did not ask to be born an Inuk
Nor did I ask to be forced
To learn an alien culture
With an alien language
But I lucked out on fate
Which I am unable to undo

I have resorted to fancy dancing
In order to survive each day
No wonder I have earned
The dubious reputation of being
The world's premier choreographer
Of distinctive dance steps
That allow me to avoid
Potential personal paranoia
On both sides of this invisible border

Sometimes this border becomes so wide
That I am unable to take another step
My feet being too far apart

When my crotch begins to tear apart
I am forced to invent
A brand new dance step
The premier choreographer
Saving the day once more

Destiny acted itself out
Deciding for me where I would come from
And what I would become

So I am left to fend for myself
Walking in two different worlds
Trying my best to make sense
Of two opposing cultures
Which are unable to integrate
Lest they swallow one another whole

Each and every day
Is a fighting day
A war of raw nerves
And to show for my efforts
I have a fair share of wins and losses

When will all this end
This senseless battle
Between my left and right foot

When will the invisible border
Cease to be

Slamming Doors

~

BY

RON

REDMOND

hach Ven ought to consider himself in good company, right up there with Albert Einstein, Philippine President Corazon Aquino and Russian writer Alexander Solzhenitsyn.

Like them, Thach has spent a good deal of his adult life as a refugee. But while Einstein, Aquino and Solzhenitsyn were welcomed with open arms by the international community, Thach has had the door slammed in his face at nearly every turn.

Nine years after fleeing Viet Nam to the sprawling Site 2 camp on the Thai-Cambodian border, Thach has yet to find a country willing to accept him for resettlement.

"I can't go home," said Thach, a member of Viet Nam's minority Khmer Krom ethnic group who had to flee in 1982 when he was accused of supporting non-communist guerrillas. "I want to be resettled in another country, but no one will take me."

It is a story heard over and over again in refugee camps and at border crossings around the world. Fearing an uncontrollable influx of people from underdeveloped nations, more and more wealthy countries are pulling in the welcome mat. Even the right to seek asylum, one of civilization's oldest and most honored principles, is under threat almost everywhere.

Thach is one of more than 17 million refugees worldwide, people who are outside their countries of origin because of well-founded fears for their lives and liberty. The vast majority of them are living in squalid camps and villages in Third World nations, the very countries which can least afford to help them. At least 15 million more people are believed internally displaced within their own countries because of civil war, famine and other disasters, most of them also in the Third World.

Like Thach, every refugee has a tragic story to tell—of persecution, of war and violence, of the loss of home, family and country.

Increasingly, it is also a story of rejection by an international community that has become weary of caring for the growing numbers of dispossessed.

In 1991, the United Nations High Commissioner for Refugees spent about $980 million caring for the world's refugees, money that many governments felt could have been better used at home. If donor governments can't take care of the homeless on their own streets, the argument goes, why should they be responsible for the world's homeless?

Making matters worse, hundreds of thousands of people from poor nations are flocking to the North and West not because of persecution, but because of poverty and the desire for a better job. These so-called economic migrants often abuse the asylum process with fraudulent claims, clogging the asylum pipeline and making it increasingly difficult for political refugees to get a fair hearing.

Despite the much-heralded promise of a new era of peace and prosperity brought on by the end of the Cold War, many of the root causes of the global refugee problem persist—regional conflicts and civil wars; human rights abuses; the growing economic disparity between North and South and East and West; government incompetence and corruption; environmental degradation; famine, drought and other natural disasters.

Add to these perennial problems the recent rise in nationalism and the disintegration of countries like Yugoslavia and the Soviet Union, and the flood of refugees threatens to turn into a virtual torrent.

To avoid these unprecedented mass movements, some experts are now calling for a reassessment of the way the international community deals with the refugee problem. They say that instead of waiting for people to be displaced, international organizations led by the United Nations should be concentrating on removing root causes—such as violations of basic human rights and extreme poverty—within the countries of origin. But doing so would require vast amounts of foreign development aid in a world of shrinking resources.

In the meantime, the rich nations of the North and West

can be expected to further tighten their immigration laws and take a much harder look at those claiming to be refugees.

Western Europe is already doing so. Although a much-feared mass migration from Eastern Europe has so far failed to materialize, wealthy European nations are struggling with hundreds of thousands of economic migrants from North Africa and the Middle East who arrived under the guise of political asylum. These migrants have tied up the already lengthy asylum process, bringing calls for speedier and more strict enforcement so those failing to qualify as refugees can be returned as soon as possible.

For all of Western Europe, the number of these asylum-seekers has risen from 14 000 a year in 1973 to 440 000 in 1990, and it continues to climb. European Community governments are especially concerned about controlling population movements after 1992, when markets and borders may be opened within the entire 12-nation bloc.

UN High Commissioner for Refugees Sadako Ogata acknowledges there may be a threat to the stability of individual nations by the unchecked mass movement of people across borders.

But she warns that the new fortress Europe mentality also poses a threat to the principle of asylum.

"As the world goes through momentous change, we are faced with a major dilemma," Mrs. Ogata said in a recent speech at the Vatican. "How do we preserve the traditions of humanitarianism and asylum, how do we treat humanely all those who arrive at our borders, without at the same time jeopardizing the safety and stability of our own societies?"

It is a question governments around the world are grappling with and so far there are very few answers.

Complicating Europe's problems is a continuing labor shortage that some demographers predict will require the importation of hundreds of thousands of foreign workers for years to come. Although their economic well-being may be dependent on these migrant workers, many Europeans are convinced that their national identities, cultures and jobs are threatened by foreigners. In several nations, this fear has given rise to increasing xenophobia and racism.

The problem is so widespread that at its December 1991 Maastrich summit, the European Council issued a declaration noting its "concern that manifestations of racism and xenophobia are steadily growing in Europe."

"The European Council... expresses its revulsion against racist sentiments and manifestations. These manifestations, including expressions of prejudice and violence against foreign immigrants and exploitation of them, are unacceptable," the declaration said.

The council urged member states to "act clearly and unambiguously to counter the growth of sentiments and manifestations of racism and xenophobia."

In Germany, the domestic counterintelligence service reported more than 100 attacks by neo-Nazis and skinheads against foreigners in 1991. In October 1991, the first anniversary of German unification was marred by hundreds of such attacks, with young thugs burning down immigrant hostels and staging angry marches demanding a Germans-only Germany.

Similar attacks on refugees have taken place in Sweden, long considered one of the most hospitable havens for the dispossessed.

In Austria, the far-right Freedom Party has ridden a wave of anti-immigrant and anti-Jewish sentiment to strong showings in three elections.

"Twenty to 30 percent of the Austrian population have negative views of foreigners, including Jews," Austria's Gallup Institute reported in November 1991.

In France, the xenophobic National Front continues to gain popularity in the polls, and even mainstream politicians are now calling for mass deportations and tighter curbs on immigration.

In Britain, the Conservative government recently announced plans to curb the entry of bogus asylum-seekers by speeding up the review process so as to reduce a backlog of more than 60 000 cases. More than 60 percent of those cases involve immigrants from Africa, and around 20 percent are from South Asia. The number of asylum applicants in Britain has grown tenfold in the last three years, to an estimated 50 000 annually.

In Italy, thousands of Albanians were forcibly returned home during the summer of 1991 despite suspicions that some may have been legitimate refugees.

Hostility toward asylum-seekers isn't confined to Europe, however.

Within hours of Italy's forced return of Albanians, for example, Malaysia claimed the right to do the same to Vietnamese boat people in its territory. The Malaysians, who came under very heavy international criticism in the 1980s when they turned away boat people, pointedly asked why the international community had remained silent when rich donor countries like Italy refused entry to asylum seekers.

In Hong Kong, tens of thousands of Vietnamese boat people found to be economic migrants face return because they do not qualify for refugee status and no other country will have them. Fewer than 20 percent of the 62 000 boat people currently languishing in Hong Kong's crowded detention centers have been able to pass a UN-approved screening process for refugees.

Hong Kong government officials have warned that the policy of first asylum—allowing would-be refugees ashore until they can be screened—may be revoked if the influx of economic migrants continues. The problem was further exacerbated early in 1991 when thousands of Vietnamese were found to be making the trip to Hong Kong solely to take advantage of a UNHCR program begun as an incentive to voluntary repatriation. The program paid $410—twice the annual salary of a Vietnamese civil servant—to anyone who requested voluntary repatriation to Viet Nam. UNHCR suspended the incentive program in September 1991.

Canada, grappling with a backlog of refugee claimants, is considering legislative changes to better screen applicants, boost the enforcement power of immigration officers and increase the ability to deport those found ineligible for refugee status. About 5000 bogus claimants were expected to be deported in 1992.

The United States, which prides itself on its heritage as a haven for the world's dispossessed and downtrodden, is accused of applying its asylum policies unevenly. While Washington criticizes Hong Kong for forcibly repatriating Vietnamese boat people, the United States has for years been doing the same thing to asylum seekers from Central America and the Caribbean.

"At least the Vietnamese can get into Hong Kong for first asylum," said Maurice Belanger, a spokesman for National Forum, a refugee advocacy group in Washington, D.C. "Haitian boat people can't even get to the U.S. coast before they're intercepted."

In November 1992, thousands of Haitians who fled their homeland in small boats following a coup against the island nation's democratically elected government were held for days aboard Coast Guard cutters while U.S. officials tried to find other countries that would take them.

The U.S. vessels returned 538 Haitians back to their country before a Miami court ordered a temporary halt to the forced repatriation.

Cheryl Little, an attorney with the Haitian Refugee Center in Miami, said Washington was trying to deny first asylum to the Haitians because of fears it would trigger a new wave of boat people.

Bill Frelick, an asylum specialist with the private U.S. Committee for Refugees in Washington, D.C., called the treatment of Haitians "blatant racism" and one more example of the disregard many Western governments now have for the principle of asylum.

Frelick said Western countries were more than willing during the Cold War to grant asylum to the trickle of refugees who managed to escape East Bloc nations.

"It was easy to grant asylum back in the Cold War days when you only had 10 Albanians a year leaving Albania," Frelick said.

"But we've clearly entered a new era where we no longer have a Cold War or the ideologically fuelled humanitarian response that it once generated."

"The trend we see now is to run roughshod over the rights of due process."

Elena

BY

PAT

MORA

My Spanish isn't enough.
I remember how I'd smile
listening to my little ones,
understanding every word they'd say,
their jokes, their songs, their plots.
　　　　Vamos a pedirle dulces a mamá. Vamos.
But that was in Mexico.
Now my children go to American high schools.
They speak English. At night they sit around
the kitchen table, laugh with one another.
I stand by the stove and feel dumb, alone.
I bought a book to learn English.
My husband frowned, drank more beer.
My oldest said, "Mamá, he doesn't want you
to be smarter than he is." I'm forty,
embarrassed at mispronouncing words,
embarrassed at the laughter of my children,
the grocer, the mailman. Sometimes I take
my English book and lock myself in the bathroom,
say the thick words softly,
for if I stop trying, I will be deaf
when my children need my help.

The Poverty of Affluence

~

BY

JACINTA

GOVEAS

efore I immigrated to Canada from Pakistan, I had often been warned about "culture shock." I had already experienced culture shock when I spent three years working as a community development worker in a village in India, more than 2400 kilometres away from my middle-class home.

Most of the people were landless farmers whose whole families, including the children, worked on land owned by absentee feudal landlords. They were paid with a share of the crop, selling most in the neighboring towns and keeping the rest for themselves. They were very poor and had very little of what we consider the necessities of life—a choice of food products, clothes, entertainment. Even the real necessities—drinking water, staple foods, medical assistance, education—were hard to come by and people often died as a result. The injustices they suffered as a result of the system of landless farming made it impossible for them to escape their way of life.

Since coming to this world, my mind has often been transported back to that village. For people such as myself who come from very restrictive cultures, North America offers the possibility of realizing so many of our hopes and ambitions.

Couples being able to demonstrate their affection on the streets, people being allowed to express different opinions and freely criticize the government, a choice of sexual preferences, the right to pursue whatever career path you want regardless of sex or class—this is what I thought made the difference between a free society and a restricted one.

I was not at all bothered by these differences. What was a real shock to me was the abundance of material goods here, from food to clothes to gadgets that we are seemingly unable to live without. Advertising campaigns make sure that we really need these things. We end up feeling apologetic because we don't have as much as our neighbors. The sense of not being acceptable was strong—and confusing. Many of us had never before experienced ourselves as different because of our accent or the color of our skin.

We may not recognize racism in our everyday lives, but we do know that what we are, who we are, is not enough. Our qualifications, our culture, our language, our food—we sense that all of this is not acceptable to the dominant community.

And so we resort to compensating for the things we feel we are not by accumulating material possessions. We need to feel accepted into the new society, so we get what they have. We decorate our houses like theirs; we use the same cosmetics, clothes, cars—whatever it takes. Rather than share the values we bring with us we bow to the obvious material values.

And in the end, we promote the myth that immigrants all come to this country for economic prosperity.

It is sad to listen to newcomers, particularly refugees, who have risked a lot to come here to join a friend, relative or even acquaintance from their home community. It is hard for the people here to find time for the newcomer. They are busy with their jobs—sometimes two jobs—paying off mortgages and loans, saving for the new car or for next year's vacation or to start a family.

I realize that all this is important, but are we losing sight of our priorities? Are we playing into the hands of the corporations by accepting their ideas of what we need? Are we agreeing to live our lives as others think we should?

My first job here was in an office. I was one of the hundreds of nine-to-fivers, but with a difference—I was the new kid. My colleagues were kind, and so decided to take my education in hand. It was interesting to listen to the things they felt I simply had to know: The kinds of clothes I should wear, the need to develop an image, an interest in makeup, all the places I simply had to go to, and how important it was to have a man in my life.

What struck me most was that they seemed to have no sense of who they were as people. All of the talk was of shopping and credit card limits. People who had worked together for years had developed very superficial relationships with one another. There was little awareness of Canada, never mind the rest of the world.

Poverty was not an alien experience to many of this group, but it was poverty within a local, narrow frame of reference: It meant not being able to afford a particular dress, feeling intimidated by the prices of electrical gadgets, having to wear the same coat for a third winter in a row. Christmas was the saddest time of all because there was so much that they wanted to get for their children and friends but could not afford.

Before I came to Canada, I used to envy Western women for their political awareness and freedom to fight for a different way of life. I took it for granted that they were all as aware as I of the possibilities available to us if we were free enough to explore them. This was one of the main reasons I wanted to make this country my home. We were socialized to think that almost every woman in the West is a disciple of women's liberation. Imagine the disillusionment when I found that women here were "liberated" only to the extent of having accepted a different set of rules, instead of having the freedom to question what was being offered.

Very few, for example, questioned the way management was using women's liberation slogans to encourage women to accept exploitation. Few had questioned the codes that made women feel they had to struggle to be more decorative rather than rise within the company by pursuing professional development that would help them to get ahead. Women, like immigrants, had accepted and integrated the values of the dominant culture.

A friend of mine, a new arrival, was talking to me about how appreciative he was of his new country. He was especially grateful to the advertising on television. "At least we know what to buy," he said.

Nobody ever said a truer word. The advertising not only tells us what to buy, it gives new meaning to old words, like "freedom for a dollar." The Lotto 6/49 advertisement proudly proclaims "freedom for a dollar" and flashes images of dreams fulfilled—fancy cars, boats, vacations to exotic places. Freedom becomes synonymous with the acquisition of possessions.

Where I came from, material possessions were few. But what we did have, and were rich in, was a deep sense of what was right and what was important. We always had time for our families and for our children. There was no need for Hallmark cards to tell our children how much we cared, because we expressed it to them every day in so many ways.

Relationships there were real. People invited you into their hearts and into their families very quickly. There was no pressure to rush around trying to meet mortgage payments or beat credit card debts, and so there was time to be present for people in a very deep way. No need to worry about counselling bills, because there were people right there to talk to, people who would be ready to walk many miles with you in your struggles.

I often think of the obvious poverty of the people in that faraway village, and of their wealth, and of the obvious wealth of people in this society. Here, there is not much you could wish for that is not obtainable—if you have the money. So we spend a lot of time and energy getting the money that is going to make getting that particular thing possible.

But are we not losing sight of something very basic in this rush for collecting material wealth? What about our relationships with one another?

I am continually struck by the loneliness and pain in this society. We feel the need to legitimize our friendships, so we have programs like Big Brothers/Big Sisters. Special committees are appointed to welcome new people into the neighborhood. What does it say about us that we need others to organize us to do things that should come naturally to us—and have come naturally to us before?

When I made the transition from my home to this country, I came looking for personal freedom. I appreciate this above all else in this society. But faced with consumerism and the need to fill in the empty spaces with things, I think we are all called upon to choose alternative lifestyles and make choices that will enable us to leave something other than our possessions to the ones who come after us.

The Earth is Weary

~

BY

SUJEETHA
SIVARAJAH

The earth is weary of our foolish wars.
Her hills and shores are shaped for lovely things,
Yet all our years are spent in bickerings
Beneath the astonished stars.

I Was Sixteen

~

BY

RAJAKUMAR

THANGARAJAH

hen I was 16, my village in Sri Lanka was attacked. At that time my mother's two brothers were shot. While our village was under military control, the guerrillas attacked the military, and in retaliation the military forces shot innocent people on the street. One day a group of Tamil liberation fighters visited our school. They encouraged us to join them and wrote down our names.

When my family heard that I had attended a few of their meetings, they were very angry. Instead of beating me, my father gave me some advice. He said that I should be a role model for my three younger brothers, and I owed it to them not to get involved in life-threatening situations. Also, since my mother was in mourning over her brothers' brutal deaths, I owed it to her to stay away from trouble.

However, I didn't listen to my father. Within one week, I left home and went to a forest with some of my school friends in order to get training from the liberation fighters. I think my curiosity and thirst for adventure took me there. When I heard that they were not going to train me to fight, but rather use me to distribute notices, I was disappointed. But in a way, it was more dangerous than fighting because we didn't have any protection. We could be shot and simply disappear.

When I returned home after three weeks, I was scared to death to confront my father. He told me, "Because of you, all of us are in danger." He forbade me to go again. He restricted me from going outside our house. But I refused to listen.

One day, after my parents had left for work, my friends came to my house with three bags of notices. They asked me to go and distribute them. I couldn't resist. After we gave out some of the notices, we were bicycling home. But the military forces caught some of the people who were reading our notices and beat them until they told our names. Within a few minutes they rounded us up, and we were trapped with nowhere to run.

They arrested us and forced us to tell who had supplied those notices to us. But we had already been trained to tell the same story over and over again. We told them we didn't know who had given them to us. But they didn't buy our lies. In a final effort to make us tell the truth, they commanded us to put our heads in front of each of the armed truck tires. They threatened to run over our heads if we did not tell the truth. I fell unconscious.

When I woke up, I was in the hospital and my father was next to my bed. Because my father was working at the hospital, he knew all the doctors. With their help I managed to escape from the hospital and then from my village to our capital city. I went to India and then to Europe and then to Canada. After arriving in Canada, I heard that two of my friends had died in battle and one was still in prison.

I was 16.

Hands of Lead,
Feet of Clay

~

BY

JAMAL

MAHJOUB

H is hands are made of lead, his feet of clay. Tiny animals live in the hardened ridges on the backs of his heels. Long sinewy loops of rope for arms, the kind you might use to string a bed with. The night is cold and hard as the strip of cardboard upon which his bones lie.

The voices are distant whispers. When he closes his eyes he hears the words of his father, telling him the same stories over and over. He has not seen his father since they came for him five years ago; crying and screaming, falling to the ground like a rag. All his learning fell from him like leaves shaken from a tree.

Now he was out in the great wide sea of nothingness, the stone-shod desert which he had watched from the open windows of the school, year in and year out until if a single stone had been moved he could have sworn he would have noticed. His father taught them. The sun, he said, does not pass over the earth from east to west, it is the earth which passes around the sun.

He dreamed of the city. Just before he fell asleep, curled like a coil in the sand he stared at the stars which seemed so close. He saw coloured lights flashing through the heavens, streaks of light so vivid that if he could just lift his arms he might be able to reach out and grasp them. The city was light, the rest of the world was darkness.

He was drawn back to the school by the radio which sang every night when the streets were deserted and his mother and sisters sat hanging their heads weeping over what was to become of them. The boys who lived in the school were the same boys who had sat beside him, or the ones from the next class, or the ones he had watched playing football as they still did in the long afternoons, running barefoot and dusty to kick a ball that was more holes than leather. The school was theirs now for all others had abandoned it to them. The teachers had packed their bags and gone, those that is who were not dismissed or arrested. The old watchman still slept against the crumbling wall, but no-one paid him anything any more, neither money nor attention.

They all knew him. They knew him by his hands of lead and his feet of clay which moulded themselves to whatever ground he walked upon. They knew him by the cartridge case that he had punched a hole through with an old nail and a stone to string around his neck and they knew him also as the son of a man who used to teach in this school in the days when it still was a school.

He was drawn in by the radio with its jittery litany of love and gibberish. When the speeches came on they switched it off to preserve the batteries, for such talk made you no wiser. They prized batteries above all else and in conducting house searches or stopping the odd lorry that rumbled lopsidedly by, the only thing they really hoped to find was batteries. He was drawn there and he stayed, hooked to that transistor like the strand of wire that was twisted around the top as an antenna.

He clung to the crippled walls where the flags fluttered, coloured rags in the soft evening breeze and he listened, for when they sat and talked the children grew wise and their eyes began to glow the way charcoal does when fanned. They were like a family, and like any family there were fights from time to time, but they never lasted long.

Once in a while a man came to see them. He was short and stubby and his eyes drooped at the sides as though he were weary of himself. He had a wristwatch that gleamed dully and skin the colour of those ochre rocks that inhabit the plains. They said he was the Government's man, that in the army he used to jump from aeroplanes, though it was hard to imagine those sorry-looking eyes ever contemplating such a daring act.

He told them that they were the future, that the ways of the past were wrong, that there was a war, a war between ignorance and knowledge, against evil and the ruin of man, against those who sought to lead them from the true path, the straight path.

With his feet of clay, his spine was like the twisted stem of a tree from which thorns burst into the glassy shimmering air. Hatred belonged to the past, his father said, there was only one God and He belonged to all men. A fool claims to know everything; it takes a wise man to admit his ignorance. But all these clever words never helped him when they came for him, all his books and papers never saved him then.

He no longer went home any more. There was really no point. He had no wish to sit there night after night in that awful air of mourning which covered his mother and sisters like ashes. They scolded him and asked him where he had been and cursed and cried and begged him to stay at home; then they slumped on the edge of the bed side by side like three ragged crows.

It was a small unremarkabe town. There were rumours of murder and slavery, of women and children being burnt alive, people fleeing for their lives. No-one paid such stories any heed for they were old stories; tales that had been told by toothless grandmothers for as long as anyone could remember. The trouble lies beyond the town in the regions still inhabited by dangerous savages who defy the Government, who poison the wells and refuse the word of God, unbelievers who still worshipped rocks, not people from this town.

From time to time there was fighting in the hills to the southeast. The stubby fellow drew them into a circle. He wore grey trousers and a white shirt that hung over his belt where his waist protruded. He held the rifle loosely in one hand, dangling by his left side. He waved them to come forwards. Watch, he said and he flipped it clumsily from one hand to the other, raising it to his hip and pulling back the bolt. He showed them the cartridges and he sighted along the barrel at the pockmarked wall. You, he said, the one with that thing around your neck. So he stepped forwards and the man handed him the gun. It was heavier than he had expected and the man laughed. How old, boy— how old are you? Thirteen, he lied, for he was tall for his age. Like this it fires one bullet, and like this it fires many, the man beamed.

The drill over, they climbed excitedly into the back of the small pickup with barely enough room for them to stand. The car rolled slowly away, bouncing over the waves dug into the ground by countless lorries and last year's rains. The desert was transformed into a wide golden sea upon which they floated. They laughed at the feel of the wind in their faces, falling against one another as the vehicle rolled from side to side. The army was busy, surprised by a sudden attack; all their forces were concentrated to the east and there was a danger now of the rebels breaking through their flank across the mountains.

The droopy fellow climbed from the cab and lit a cigarette. There was a small dry valley which turned into a river during the rainy season but was dry as a bone now. Its sides were flanked with stunted trees and patches of yellow grass. Follow that up towards the top and stay there by the sandy grey rocks you can see. He walked along the troupe of boys inspecting them. He paused, his eye drawn by the flash of light on metal. He reached out a hand for the brass case. Why stuff the neck with cloth? he asked. To keep the rain out, the boy answered. The others laughed, but the man looked concerned. Amulets, talismen, superstition. When would these people give up their old beliefs? He turned away in irritation and waved them on their way.

They started up the valley together, but soon he had put a distance between himself and the others. They stopped to chat, to smoke cigarettes and look at their guns. His bare feet left no trace on the hard ground and soon he could no longer see them when he looked back. He slowed a bit. These mountains were strange. The air was different and the sky was deeper, more blue. It felt like a drum skin against which he rested his head. A bird flew over and he raised his gun, sighting along the barrel. He liked the feel of it in his hands. The sides were scratched and beaten. The wood which he gripped was worn smooth by other fingers. It surprised him how well it fitted to his hands, as though all of those people who had handled it before had been preparing it for him.

It was so still and silent. He had no idea of time, of how long he had been walking. The sun was falling, crashing towards the plain. He sat down to watch as the shadows grew longer. His father was wrong, the world was flat. When he turned his head he caught sight of a small lizard on the ground beside him. Stripes of orange and green ran the length of it. Its belly rose and fell at a fast beat. He stretched out a hand and it was gone. He continued upwards towards a pile of sandy grey rocks that seemed to have been stacked there by giants.

The man must have been crouched there among the rocks for he suddenly appeared, rising up before him. The tallest man in the world with long sharp limbs, bones like a bird. The two of them stood there, silent and motionless. Then the man's taut face relaxed and he seemed to say something, a greeting perhaps or a question. A great flower bloomed on his chest, spreading its crimson petals across the khaki fatigues. He vanished from sight as suddenly as he had come.

The sound of the shot came slowly as though off in time, many hours, many days away. He looked down at his hands and marvelled at the way in which his fingers fitted to the wood. The way the metal seemed charged. The ground beneath his feet was crawling with ants. He felt his muscles growing in his arms as he grew towards the rifle.

His hands are made of lead and his feet of clay. He remains standing there unmoving as the sun becomes the moon and his shadow turns silver and faint. When he reaches the bottom of the hill the laughing boys are gone, there is no-one about. His father was a fool. All that nonsense about ghosts and spirits being a thing of the past when nothing in the world had changed. It was the books and the papers with their mystical signs which had confounded him. The hills still rattled to the tune of their ancestors' dead bones. He begins to walk. Night becomes day and day night and so it goes. The sun blisters his tongue. His eyes are two dark holes torn in the blue blue sky by the hawks escaping through to the dark extraordinary night beyond.

The 1988 Nobel Peace Medal,
awarded that year to UN Peace-Keeping Forces

Love and Death in Sarajevo

BY

CRISELDA YABES

Marija held my arm reassuringly. "And now," she said, holding her breath, "the adventure begins. Are you ready?"

I had no other choice.

We were leaving the damaged newspaper building in Sarajevo where she works. There was no other way to get out except to cross a deserted railway line exposed to sniper fire.

Marija was still holding my arm. Her friend Senad was our front guard, armed only with an umbrella. We dashed across the tracks, took a safer route behind an abandoned warehouse for dilapidated trucks, them crawled into the bushes beside a river. That over, we faced the final obstacle—climbing over a fence; if done quickly it meant we had less of a chance of being shot.

For a split second I was so immobilized by fear that Marija and Senad had to pull me off the ground. But we made it to safety. Marija gave me a hug of relief and we broke into nervous laughter, "I think we need a vacation in Los Angeles," Senad said jokingly, "where the sun shines and there are people everywhere."

Almost every day since the civil war came to Sarajevo ten months ago Marija and Senad have taken the same route to and from work. They've naturally developed the mental strength to ward off fear— the bravery of fatalism.

Marija, a reporter, and Senad, a photographer with the underground newspaper *Oslobondenje*—which means "liberation"—have taken upon themselves the dangerous work of covering a complicated war.

That sunny day Marija was wearing a flak jacket for the first time. A year ago before she became a war reporter Marija had a glamorous job working for a women's magazine. She hadn't lost any of her sophisication. Her lips were painted red, her hair nicely trimmed. She wore a fake diamond stud in her left ear, a dangling earring in her right and a bright blue jacket over her bullet-proof vest.

Everywhere I went in the former Yugoslavia the images of women like Marija belied the ugliness of war. It was as if keeping themselves beautiful with cosmetics and stylish clothes were a psychological defiance of misery. "We have to do this even if war is going on," she told me. "We can't just stay in the basement. Life has to go on."

Marija left me at the guard post of the UN military base that was once a post office. She reminded me about the letter she had written hastily over lunch. It was a love letter I was to deliver to a man named Amer in Pristina, where I would be going the following week. I promised her it would get to him. We embraced each other and said goodbye.

As I walked into the UN base a man with whom I spent most of my three days in Sarajevo waved at me. Nebosja "Neso" Marijanovic beamed an avuncular smile and pulled me close to him. "How's my favourite?" he said, as if I had just come home from the neighbourhood playground.

Neso is a Serb. Anyone who thinks Serbs are monsters would not think it of him.

Neso is as much a victim as any Muslim or Croat of the civil strife in Bosnia-Hercegovina. His wife is a Croat; he had to send her and their son to a refugee camp in the Croatian city of Split.

For hours he talked to me about his life "before all this." Those who tell their stories steer away from the tragic narratives that are the stuff of journalism. He sticks to the past: "I was an economics major working for a computer company." And to the present: "I'm now a driver for UNICEF." He takes the emotional leap to avoid what happened in between—to his own people. Pointing to the urchins lurking outside the fence of the UN base he said: "They used to be bright boys who went to school every day. Now they're beggars."

Neso showed me what has happened to Sarajevo. In a white Land Rover he drove me through the streets of what had been one of Yugoslavia's most charming cities. The park had lost its habitués; the presidential office stood as helpless as its people; the cemetery was overcrowded. Near the cemetery is the stadium, now a storage space for humanitarian relief goods, once the pride of Sarajevo during the

1984 Winter Olympics.

Neso counts himself lucky. Working for a UN agancy at the heavily guarded military head-quarters of the UN Protection Force (UNPROFOR) accords him the privilege of electricity, water and heating—the basic necessi-ties of which many Bosnians are deprived. He has access to armoured personnel carriers (APCs)—the only mode of trans-port, aptly called "taxis"—and to the daily flights of about 15 UN cargo planes known as "Maybe Airlines," which may or may not give you a lift out of Sarajevo once they've unloaded crates of humanitarian food.

"This is California," said Neso, gesturing to a four-story building with a basement. "Outside is Cambodia."

I was given a room in the "California" basement with French nurses. It is the safest place to be. The daily grind of military planning to help human-itarian relief operations takes place upstairs. That's where I ran into the captain.

On our first encounter he took me to the rooftop, from where we could see the deadly "fireworks" at a distance and hear the sound of artillery amid the darkness of the city. "The Serbs fire 5000 artillery shells every day," he told me. "People are still living in that apartment building." he added, pointing to the one closest to the firing. "They sleep at night but they don't know if they will wake up the next day."

Inside the captain's office another UN officer appeared to join us in conversation. He car-ried a thick file of papers and dropped them on the floor. They were records of Serb prisoners in Muslim hideouts: 700 in Tarcin, 1050 in Pazaric, 23 in Hrasnica (including 16 of them beaten to death); and Muslims in Serb camps: 57 in Kula, 479 in Hadzici. The officer had collected dozens of reports about prison camps from both sides.

A soldier jaded by war can either pretend it doesn't exist or joke about it. This officer had the gift of humour and he turned his experiences into a comic adven-ture that lasted all night. We were within listening distance of the commander's quarters and we had to lower our voices like adolescents disobeying a curfew.

I asked if the senior UN offi-cials had taken any action con-cerning the reports.

"It's buried underground," the officer told me.

"Do you know what that means?" asked the captain. I nodded. It meant the papers had been consigned to an obscure corner, waiting to gather dust on a shelf.

Politicians arriving regularly in Yugoslavia for "assessment trips" leave without answers. Workers on the ground focus on short-term solutions, like handing out food to refugees.

"I'm solving hundreds of little problems every day but I don't have the overall solution to the problem," said Anthony Land, head of the office of the UN High Commissioner for Refugees (UNHCR) in Sarajevo. "It's a humanitarian problem, not a humanitarian solution," he said. "We're buying time for the politicians to come up with a solution." And if there is no solution in sight? "We'll buy more time."

Colonel Robert Bresse says there is a solution.

"The Serbs need a beating. We need to have planes up in the air and hit targets. If we don't interfere we will be here for 10 years."

This is a French commander talking.

The ramrod Colonel Bresse, who is chief of the UNPROFOR infantry battalion in the Bihac Muslim "pocket" says things that could get him into trouble with his superiors in Geneva. He says he doesn't care.

He holds court at his headquarters. His men adore him. He regularly entertains visitors and journalists in the officers' mess, with soldiers in uniform serving a wide selection of cheeses, St. Emilion wine and roast chicken or hare. The chef must never run out of chocolates—they are the colonel's favourite.

He barely concealed his sympathy for the Muslims—or his disdain for the Serbs. He kept such thoughts to himself, however, when he came face-to-face with a Serb colonel. At a pre-arranged meeting in the abandoned Hotel Kamesko in Donji Lapac, Colonel Dusan Banjac was pleading for humanitarian help.

Colonel Bresse showed up at the rendezvous with Jacques Franquin, the outspoken Belgian chief officer of UNHCR in Bihac. The Serb colonel, seeing that Franquin brought his female Muslim translator with him, remarked: "Ahh, why do you have such a pretty one and I get stuck with this?" jabbing a finger into the underling who served as his interpreter.

Franquin's translator said nothing and sat down to begin a gruelling two-hour session of translating from Serbo-Croat into English. The Serb colonel is a man she holds responsible for wounding her father and the death of her best friend. Occasionally she smoked, holding her cigarette with trembling fingers.

Colonel Bresse privately believed the food would go to Colonel Banjac's ill-fed troops. Franquin therefore gave the Serb a hard time. He demanded permission to visit every house and meet with civilian authorities. He said that if the Serbs wanted access to food they must open a restricted road leading to Kajina.

Colonel Banjac scratched his head and pounded on the table. He would agree to every demand except opening up the road.

"Then maybe some kind of a cease-fire can be arranged," Franquin suggested.

The Serb was astounded.

"If there were a cease-fire," he replied, raising his voice, "then there would be no war!"

The look of disgust on Franquin's face was unmistakable. Alone with Colonel Bresse after the Serb officer had left, Franquin blurted out: "It was all bull—!" The French colonel tried to lighten Franquin's mood. He said teasingly, "I think this is the beginning of a love story."

There are real love stories, even in wartime.

I kept my promise to deliver Marija's letter to the man she had not seen in ten months.

When I arrived in Pristina I immediately called Amer. He came right away to get the letter and returned to my hotel the next day with a reply for Marija.

Amer has a handsome face, and it dissolved in fear and self-pity as he spoke. He was close to tears while recounting the first days of war in Sarajevo. A Serb had aimed a rifle at his face. That near-death experience had driven him away from the city where he had lived for a decade and had fallen in love with Marija. He had to leave because he is a Muslim. Marija would not come with him. She had to be with her family. Her father is a Croat and her mother is a Serb. She calls herself a "Bosnian."

"You're a journalist," Amer said, speaking to me in a flat voice. "You can look at war, but you can never feel it in your skin." Amer used to be a journalist too. He is now a construction worker, saving money to seek asylum somewhere, anywhere— "maybe Germany, maybe Switzerland, maybe Austria"—to escape the war.

"I told her she must live, not because I want her to live," he said, putting the letter in my hands. "She must live because she must never give the Serbs the glory of killing her. And maybe...maybe years from now we will be together again."

Synonyms
for
War-torn

~

BY

OAKLAND

ROSS

has Whepler got a call from a friend of his who worked at the Catholic human-rights office. The woman said about six children had gone missing that morning from a *barrio* over by the local cement factory. She told him how to get to the *barrio*. She said she thought it was six children. She wasn't exactly sure about the number.

Whepler grabbed his notebook and rode the elevator downstairs. He didn't want to be doing this, but what choice did he have? Come to that, he didn't want to be in El Salvador anymore. He was thinking of getting out, thinking about it a lot. In the lobby of the hotel, he bumped into Jean-Marc Piton, who was flirting with the girls behind the reception desk.

Jean-Marc looked up, threw out his arms. "Charles!" he cried. He hurried over. He said he had a big problem. The French ambassador in Belize was coming to El Salvador on a sort of courtesy call, and he, Jean-Marc, was responsible for organizing a dinner in the man's honour. He'd just got the word in a telex from Belmopan, the capital of Belize.

Whepler shrugged. "So...?" He knew that there was no French ambassador in El Salvador itself. Paris had called him home. "What's the problem?"

Jean-Marc stood back, and his eyes grew wide. "But nobody in this country will come to such a dinner. Nobody." He smacked a fleshy hand against his brow and shook his head. "It will be a *désastre*, and I will be the one to blame."

Whepler pursed his lips. Probably, Jean-Marc was right. France had a socialist government, after all. Not too popular around here. "Well, do the best you can."

"You have to help me, Charles." Jean-Marc clutched at Whepler's arm. "Come. We go to the bar."

"Not right now." Whepler pulled free. "Sorry. I have to run."

"All right. Go. I will drink alone. What difference does it make? My life is over."

Whepler left the hotel and walked across the street to the Metro-centro, the biggest shopping complex in Central America. He went up to a lone taxi parked at the stand by the curb. The driver's name was Roberto. Whepler knew him from previous trips.

"*Hola*, Roberto," Whepler climbed in the front. "*Vamanos a la fábrica de cemento.*"

"*Muy bien.*" Roberto started the engine. He did a U-turn at the traffic lights, and off they went. His taxi needed a new muffler.

Whepler leaned his elbow out the window and tried to come up with some new synonyms for war-torn. He wondered what sort of adjectives you would use to describe a country where six kids went missing from the same neighbourhood on the same morning, out of the blue. War-torn didn't seem to cover it. Maybe war-crazed.

The *barrio* was down a wooded gully, and the only way in was along a dirt track that ran over a sort of garbage dump and then across a stream. First, though, they had to drive past a black Ford sedan that was parked in the long yellow grass at the edge of the track. Four men were hunched inside, listening to a soccer game on the radio. Someone had just scored, and the announcer cried out, "*¡Gooooool!*" Probably the wrong side was ahead, because none of the four men was cheering.

They all had on aviator sunglasses and were evidently police, although they weren't wearing uniforms and they didn't show their guns. They just sat there and stared at the taxi as Roberto eased by. Whepler figured they'd probably make some trouble on the way out. Maybe serious. Maybe not. He released a long breath. God, he was getting tired of this stuff.

Good for Roberto, though. He didn't blink an eye. He just drove straight on. The car crept down and across the stream and came up behind a woman who was walking beneath a stand of eucalyptus trees. Whepler asked Roberto to stop just ahead of her and he got out. He didn't approach the woman—he imagined she might be a bit jumpy. He kept his distance, on the other side of the car. He said, "*Buenas tardes,* Señora."

"*Muy buenas tardes,*" she replied. She put up her hand to shield her eyes from the sun. A roly-poly woman, she had dark skin and thick black hair pulled straight back and tied behind her neck. She wore a short blue dress with a tattered yellow apron tied at her waist. She carried a white towel over her shoulder, the way they did. She looked to be in her mid-thirties.

They exchanged a few pleasantries, and Whepler said he was a reporter. He said he wanted to find out about these children who'd gone missing.

Just like that, she started to cry. It turned out that she was the mother of one of the children who'd disappeared. Her daughter was named Alma. Just twelve years old. Whepler went over to her, and he and Roberto drove the woman to her matchbox adobe house, a tired affair with a smoke-charred roof.

The woman asked Whepler inside. She said her name was Señora Nevares. She had three other children, all younger than Alma. She said she had a husband, too, but he was out in La Libertad working as a day labourer in the cane fields—it was the harvest season. The house was run-down but tidy—a couple of dark rooms, one for cooking and one for sleeping. The chairs were homemade and creaked when you sat in them. The woman had some water in a pitcher and she got up to pour a glass.

"It's been boiled," she said. "You can drink." She wasn't crying any more, just sniffling a bit with her towel bunched up in her hands. She watched Whepler look around at the inside of her house. She pressed her lips together. "It's very small," she said.

"Yes. But it's nice."

She shrugged.

There'd been a time when Whepler thought the consolation about poverty was that poor people didn't realize they were poor. They didn't know any better, so maybe it didn't seem so bad, the way they lived. He thought they were probably accustomed to septic water, for instance, and could drink it without much effect— it didn't make them ill. But he'd changed his mind before too long. They knew they were poor. They did get sick.

Whepler pulled out his notepad, and Señora Nevares told him what had happened. Some men had come by in a pair of cars, she said. They'd stopped by the garbage dump and climbed over some mounds of trash. There was a flattened section of land there where the local boys sometimes played pick-up soccer. The men started calling for Joaquín Díaz. He was a boy, a bit older, about seventeen maybe, who lived in the *barrio*. They took Joaquín away in one of the cars. That was yesterday afternoon.

"This morning," the woman said, "they came back. The same men or maybe someone different. I don't know." They went from house to house, shouting out the names—the names of children who lived in the *barrio*. They had rifles, and they waved them in the air. As soon as the armed men found a youngster who matched one of the names, they dragged that child out and pushed him or her into one of the cars.

"How many?" said Whepler. "How many did they take?"

Señora Nevares had her head lowered. She was crying into that towel of hers. But she answered his question. "Seven," she said. Her voice was muffled by the towel.

Whepler frowned at his notepad. He was thinking about those men in that Ford up above the *barrio*. To get out of this place, he was still going to have to make his way past them. He gritted his teeth—he'd had it with this stuff. After a while, he stood up and went outside with Señora Nevares to visit some more of the parents. The other people emerged from their houses to talk. They were afraid—Whepler could see it in the tautness of their jaw muscles and the way their eyes shifted to the side. But they opened right up, and he knew that here was a problem. These people thought he was going to be able to help them in some way. They imagined he had some sort of power, some knowledge or influence, something beyond their grasp—whereas the truth was, all he had was a pretty good story for the Monday paper.

Whepler walked back with Señora Nevares to her house, where Roberto was still waiting beside the car. The woman bustled indoors. When she reappeared, she was carrying something—a small black and white photograph of her child. It looked like a school snapshot. The girl was a sweet-looking kid with thick, arched eyebrows, a button nose, and her hair cut short and wavy. She looked like the sort of child any parent would be happy to see their daughter bring home from school, a new friend. Whepler started to return the photograph.

"No." Señora Nevares put up her hands with the towel draped between them. "You keep it. Please. Maybe it will help you."

So there it was again.

Whepler hesitated, but then he tucked the snapshot into his money belt. It felt like some sort of IOU he'd never repay. He thought he should offer something in return and so he gave Señora Nevares his card. If she heard anything new, he said, she should try to get in touch.

The woman peered down at the card and nodded. Then she slid it into the front of her dress. She looked up, grim-faced, as if they'd just made a secret pact. "*Gracias.*"

Whepler turned to Roberto. "Okay," he said. "*Vamanos.*" He wanted to get going. He was feeling edgy, thinking about those four men in that Ford up on the brow of the hill.

When they got there, Roberto had to stop. The police had eased their car over into the middle of the track, to block the exit. Whepler closed his eyes, as if somehow he could imagine all of this away. But, when he opened his eyes again, he was still in El Salvador.

The doors of the Ford swung wide, and the men climbed out. The gravel crunched beneath the heels of their smart black shoes. It was late afternoon now, and the sun was behind the Ford, sharp and low, casting thin, elongated shadows. The men ordered Whepler and Roberto to get out of the taxi. A couple of them started to search it top to bottom. The other two questioned Whepler, and one kept prodding him in the chest with his hand. Every question, another shove. He got Whepler pressed with his back against the Ford, and now both these men stood close. They had shiny, bronze complexions, and their sunglasses flashed in the light. They wanted to know Whepler's name, his nationality, his occupation. They wanted to see his documents, but Whepler couldn't move to take them out. The door was open behind him, and it would not have taken much for them to push him right in.

Whepler couldn't speak. One of the men kept his hand where it was and brought his face close, his burnished brown skin. He stared through his dark-green lenses, like insect eyes. There was tobacco on his breath, stains on his teeth. His nose and cheeks were pitted, sprouting dark hairs. He was looking for fear, searching for it in Whepler's eyes. When he found what he wanted, which he was sure to do, he might go right out of control. But, instead, he just smiled and slowly removed the weight of his hand. All he'd intended to do was show who was the boss, and he'd done that now.

He stood back a pace. His manner changed at once. "You must understand," he said. "We are concerned for your safety. You have to be so careful in this country. There are so many crazy people." He gestured down towards the *barrio*. "You shouldn't believe what crazy people tell you." He shrugged. "Now, you may go."

When they'd driven a certain distance away, Whepler turned to Roberto. He wasn't sure he could trust his voice. He said, "Scared? Did that get you scared?"

Roberto rolled his eyes. "What do you think?" He held up his right hand. His hand was still shaking.

Whepler did the same and found his hand was shaking, too. They both started to laugh, and Whepler felt a backwash of weakness sag through his chest. He thought, war-weary—a new synonym, one he hadn't thought of before. He was thinking, he'd had enough of this.

Back at the hotel, Whepler wrote his story and sent it off on the hotel telex machine. From his room, he phoned the paper to see if they had any questions. Bob said nope, the story was fine. Whepler went down to the lobby bar, but it was empty. So he headed out to dinner alone. He walked through the darkness and a cool breeze, down the street to a place called El Rancho. A few of the other hacks were there. They all stayed on after dinner, had a couple more beers.

At breakfast the next morning, Grant Lovsted slouched into the coffee shop and flopped down at Whepler's table. "*Café*," he groaned to the waitress. "*Por favor.*" Some of the others were up at the buffet, spearing things onto their plates. It was seven o'clock. Lovsted rubbed his eyes with the heels of his palms and yawned. "Hey," he said. "D'you file yesterday?"

Whepler shrugged. "Yeah."

"What was your lead?"

Whepler flapped his napkin, to get rid of the crumbs. He shrugged again. "I dunno. Seven kids go missing from impoverished *barrio* in war-weary El Salvador."

Lovsted rubbed some sleep from his eye. "War-weary, eh...?" He nodded. "Not bad. I'll have to remember that one."

It was a game they played. Editors wouldn't let them use the word war-torn any more. It was too clichéd. So they always had to come up with something else. They weren't allowed to quote taxi drivers, either—it gave the impression they spent all their time in a car. These were the two main rules of war coverage: synonyms for war-torn, no taxi-driver quotes. They still did quote taxi drivers, though—called them small businessmen.

Whepler crossed his legs at the ankles. "Did Jean-Marc tell you about this dinner he has to organize?"

"All he talked about." Lovsted shook his head. "What's the French ambassador want to come here for? They don't even have diplomatic relations."

Whepler nodded, sipped his coffee. It was true. Jean-Marc was the only French guy left. He was the commercial rep, so that made sense. Couldn't stop trade.

Lovsted drained his mug. "What're you doing today?"

"I don't know. I haven't worked it out yet."

"Maybe we could take a drive. San Vicente or something. Some guys said there was fighting out there yesterday. Could still be good."

"Okay. Sounds okay." He was thinking, war-battered. War-scarred. Shell-shocked.

"Caffeine." Lovsted shoved himself to his feet and picked up his mug. "I need more caffeine." He headed for the buffet.

Whepler signed his bill and walked out into the lobby—and there was Señora Nevares with her three kids. They were propped up like worn raggedy dolls on one of the big brown leather couches, beside a potted plant.

"*Buenos días.*" Whepler went over to them. "*¿Qué pasó?*"

Señora Nevares was holding his business card in her lap. She frowned down at it. "*Buenos días,* Señor...Wayplayer."

Whepler pushed a chair over and sat down. Again he asked, what was up?

It turned out, not much. The four men in that car had gone away some time after dark and hadn't shown up again in the morning. That was about it. Señora Nevares wanted to know if he'd heard anything new about her daughter.

He shook his head. "No. I haven't heard anything."

"Oh." She looked down again and nodded.

Her kids were keeping perfectly quiet on the couch beside her, two girls and a boy. The elder of the girls seemed to be maintaining a watch on the other two. She reached over and started probing her brother's hair for lice.

Whepler wanted to think about something else, such as how Jean-Marc was going to get anybody to attend that dinner of his for the French ambassador. Or maybe not about that. What he really wanted to think about was something else entirely—getting out of this country, calling it quits. He could call his paper today, before lunch, and announce that he'd had it. After three years down here, he wanted out. He could be gone in a week. He was war-fed-up. He looked at Señora Nevares. "You never know," he said. "You never know what might come up."

"*Sí.*" She nodded. "*Es la verdad.*"

It was the truth. They both just sat where they were and thought about that. Whepler was waiting for Señora Nevares to leave. But she didn't make a move. It took him a while to realize why—she had nowhere else to go. In her search for her daughter, he was it. Who else could she get in touch with? The government? The police?

"Hey, Whepler. You ready? Let's go." It was Lovsted. He was tossing his car keys with one hand,

Whepler glanced back at Señora Nevares. "Well, I'll let you know if I hear anything." He stood up. "Don't worry. I'm sure everything will be all right." He couldn't believe he was telling her this. It was stupid, cruel even.

Again, she nodded. "*Gracias,*" she said. She and her elder daughter started to get the kids up and going. "*Muchas gracias*, Señor Wayplayer." She knotted the smaller girl onto her back with that big towel of hers, and the older girl did the same with the boy. They tottered out of the hotel on their short brown legs, in their worn rayon dresses and their flapping plastic sandals. Everyone turned to watch them go....

It turned out the rebels had attacked a garrison town in San Vicente province. Lovsted had got the basic information from Larry Schuyler, the AP guy on the second floor at the hotel. They drove out there and soon they were crawling around in the dirt between rows of small adobe houses with bright stucco fronts, trying to get as close as they could. *Crack, crack, crack* went the gunshots. The air felt hot and raw. Lovsted yanked off his white polo shirt and tied it to the end of a stick he'd found.

They crept around like that, with the white flag suspended in front. They were trying to get a fix on things, figure out what was going on. For a time, they got pinned down at a gas station. They'd been scurrying across the street when shooting broke out—and there'd been nowhere else to hide. They had to lie down right behind the pumps, a stupid place to take shelter, but there was no choice. The bullets hurtled and smashed against the low tin roofs of some shacks right behind them. There must have been rebels back there, but they weren't returning fire, or not yet. Whepler could see the top of Lovsted's head, little flakes of dried tar mixed in with the dense brown hair. "Now what?" he hissed.

"Stay down," Lovsted shouted. "And don't smoke, either."

Whepler wanted to do something, anything, to get them out of here. He couldn't stand this being so helpless, but it was no use. There was nothing to do but wait until the shooting finally moved off somewhere else, and then he followed Lovsted. They scrambled away from the pumps over towards some shabby market stalls and an adobe wall. They both slumped down there. Rifle shots snapped like whip cracks, a block or so away, maybe two blocks. Now and then, the boom of an exploding mortar rumbled through the earth, and Whepler felt it in his groin like a contacting fist.

Lovsted slapped some of the muck from his shirt so that it would look halfway like a white flag again. He glanced up. "Come on. We need to find an officer. We don't have any officer quotes."

They did find an officer, too—a cocky young lieutenant jabbering into a radio behind a sandbag barricade. The barricade was set up against the wall of a building—a bodega or something—and overlooked a small square, planted with cypress trees and some scraggly lemons. The rebels and the army were shooting across the square.

Lovsted and Whepler waited until they caught the lieutenant's eye. Then they skittered along an alley and ducked down behind that barricade, right beside him. When the officer got off the radio, Lovsted asked what was going on.

"*Combate,*" the lieutenant shouted. "*Mucho combate.*"

Lovsted shook his head. "Ask a stupid question..."

The rebels didn't appear to have an angle on this barricade, so it was possible to peer over the top and see what was happening in the square. Not much, it seemed, apart from flying bullets. Then Whepler saw something. At first he thought it was a soldier, a dead soldier, lying on his chest on the cobbles. No—not lying. He was moving. He looked like he was trying to crawl out of the square. And he wasn't a soldier, either. He was far too small, and he wasn't wearing a uniform. He was barefoot, and he had on brown pants and what looked like a blue T-shirt. He was just a kid.

What was a boy doing out there? He looked about ten, maybe

twelve years old, and he was on his stomach. Whepler saw now that he had a large, dark stain on the back of his shirt—blood. He'd been shot. He could see a long, broken streak of blood from the point where he'd evidently been hit, over by a little fountain in the middle of the square.

Whepler wanted to cry out. But what would be the point? That boy wouldn't hear him anyway. So he hunched his shoulders and just watched. The boy was trying to get out of the square, but it was slow going. He dragged himself forward a foot or so with his hands and forearms, before slacking off to rest. Then he pulled himself forward a little more.

"Can't you do something?" Whepler shouted at the lieutenant. "Can't you get in there and bring him out? He's just a kid."

"You crazy?" the lieutenant roared back. "He's dead. He'll be dead soon, anyway." The officer dropped down onto his knees and started barking again into the radio. It turned out he was calling in mortar rounds against the rebels.

"Better get down," Lovsted shouted.

But Whepler couldn't. He couldn't take his eyes off this one young boy, who'd somehow got himself caught all alone in this square at just the wrong moment. Probably, he'd been running across it, thinking the way was clear. Stupid little idiot, to get caught out in the open.

Once or twice, Whepler thought he almost caught the boy's eye, but probably not. The child probably wasn't thinking of anything or anyone but himself just now. There he was, out on the sagging cobbles of that square, with the war busting a gut just above his head, and he was all alone. Nobody else was even looking his way.

Whepler had the idea that if he just kept watching, just kept his eyes peeled, then maybe that kid would be okay. He'd crawl out of there alive. Maybe there was some kind of karma, some kind of good-luck force, in having somebody keep their eyes on you. At least, you wouldn't be alone. Whepler gritted his teeth and clenched his fists and tried to will that boy across the square to safety. He kept watching until the first of the mortars exploded, too short, and dust and bits of wood and brick clattered down. After that, Lovsted grabbed him and hauled him to his knees. He tried to get up but Lovsted wouldn't have it, climbed right on top of him.

"You crazy?" Lovsted shouted. Another mortar exploded, and more debris rained down.

Whepler flailed out with his arms, but it was no use. Lovsted was too damn big.

Before long, the lieutenant was back shouting into the radio, and three or four more mortar rounds crashed and shuddered somewhere ahead, at the side of the square. Whepler's teeth shook as though they'd broken loose from his gums. Finally the shooting seemed to quiet down. Lovsted got off him, and Whepler struggled to his feet. He peered out over the barricade.

There he was. He hadn't got far, hadn't really moved at all, and he wasn't moving now. He sprawled on the cobbles with his arms pinned underneath him and with a pale grey dusting on his backside. His trousers were torn at the side. His mouth hung open. He lay still. Whepler was pretty sure that he was dead.

Right away, the shooting started up again a few blocks farther along. The lieutenant took Lovsted and Whepler back a short distance, into a dark one-room building that seemed to be some sort of secretarial school. It had old typewriters set up on all the desks. The officer rattled off some numbers about the dead and wounded. He also provided some passable quotes about the grimness of war. After a while, the fighting seemed to calm down for a bit.

At first, Whepler didn't want to leave. He wanted to go after that young boy, look at his body. He thought maybe he should try to pay a visit to that boy's family, tell his mother what had happened to her son. But it was impossible. He didn't even know the kid's name. So he just kept low and darted through the streets and lanes behind Lovsted. They finally got to the car.

Lovsted started the engine and pulled out, headed straight back to the capital. "Whoo!" he shouted.

Once they were on the Pan-American Highway, Lovsted's foot hit the floor, and he kept it there. He still didn't have his shirt on, and he was covered all over in grease and dirt.

"Whoo!" Lovsted shouted again. The windows were open, and the wind blurred his hair. He was having some kind of adrenaline fit. "Whoo! Whoo!" He kept punching the wheel, shaking his head, and laughing. "Whoo!"

"Yeah, yeah," Whepler muttered. He popped open the glove compartment and started rooting around in there for a pack of cigarettes. He was thinking, war-wired. The guy was war-wired. Whepler got out a cigarette, but the car lighter didn't work. He had to poke around in his money belt for a book of matches. Instead, he came up with that picture, the one of the little girl—Alma.

"What's that?" Lovsted reached over and took the snap, glanced at it, and handed it back. "Cute kid. Yours?"

"Yeah," said Whepler. "Sure thing. My illegitimate daughter."

"Take good care of her, pal." Lovsted hit the brakes, and the car shook through a patch of dirt and washboard. He wailed out an old Bobby Vee refrain—"Take good ca-are of my ba-aby!" He punched the wheel again. "Whoo!"

"Yeah." Whepler slid the photo back into the pouch and got out some matches, lit his cigarette. He had to be careful, because of the way his hands were shaking. "Yeah, yeah, yeah."

The paper loved the story, though.

"Fantastic!" said Bob on the phone. "Great stuff! Great colour!" The story was going front for sure. "Keep your head down," he said. "It sounds pretty hairy down there."

"Yeah," Whepler said. "Yeah, thanks." He had to admit, though, that the praise didn't hurt too much.

Whepler went downstairs and met Lovsted in the lobby bar. Lovsted said his story was going front, too. Whepler said it was ridiculous. It was stupid.

"What is? What are you talking about?"

"This." Whepler dug a cigarette out of Lovsted's pack and lit up. "This whole thing. What did we do today? We went out and got shot at, saw some other people getting shot at. Big deal. It doesn't mean anything. We write about it, and nothing happens. What's the point? The people still die. The war just keeps on going."

"Damn right." Lovsted drained his beer and slammed down the mug. He asked Pedro for another.

Whepler shook his head. It was ridiculous. He sipped his beer and tried to think of some more synonyms for war-torn, but it was no use. He'd run out.

"Ah! *Mes amis*!" It was Jean-Marc. He waddled into the bar from the lobby, climbed onto a stool beside Lovsted. He shook his head. "It's terrible," he said. "Terrible."

"What?" said Lovsted. "Don't tell me. Still no takers for the Ambassador's Ball?"

"You laugh," said Jean-Marc. He was quiet for a time. "But it isn't funny. I am finished. Destroyed."

"Looks like it," said Lovsted. "Too bad."

Next morning, Whepler wasn't out of bed yet when he got a call from Manuel, the concierge in the lobby. Manuel said there was a woman at the hotel entrance who needed to talk to him.

When he got downstairs, Señora Nevares was out in front, with her three kids. She was perched on a low stone wall just across from the hotel entrance, under the flame trees. She was surrounded by several grimy, barefoot boys, the ones who came by each morning to sell newspapers to the hotel guests. They roughhoused and squabbled around her. Whepler walked over. Señora Nevares was full of apologies for bothering him. Whepler said it was all right, no problem. What was up? Any news?

"Nothing," she said and shrugged.

It was an exact repeat of their conversation the previous morning. She wanted to know if Whepler had heard anything. Whepler said no, not a word. He said he was sorry about it.

She nodded. She said she was sorry too.

They stayed like that for a minute or more, both of them staring off into space. It was another beautiful morning, and the traffic lurched and growled past along the Boulevard los Héroes. Señora Nevares reached over and straightened her elder daughter's dress. Whepler was trying to think of something he could say, but he wasn't coming up with much. What was he supposed to do? It was awful—but no one was going to find her daughter again. She was gone. That was it.

Maybe that was what he should have said, but somehow he couldn't say it. He was thinking, what the hell. Maybe he could go around to some government offices, ask some questions, get a lot of blank stares and curt denials, even a threat or two. But at least he'd have done something. He'd have something to tell this woman, some activity to report. Señora Nevares would have the feeling that something was being done about her girl. Whepler figured he could at least do that.

"Look," he said. "Why don't you come back tomorrow morning? I'll see what I can do. Maybe I'll have some information for you by then."

She didn't say anything. She just closed her eyes and fumbled with her towel. Whepler thought she was going to start to cry. But she didn't. Instead, she just looked up at him. She said, "Gracias. Gracias, Señor..."

"Wayplayer," he said.

"Si. Perdóname. Muchas gracias, Señor Wayplayer."

And off she went, with the kids in tow. Whepler stood there and watched them—Señora Nevares with one daughter clinging to her back, the other daughter carrying the boy. They looked like little elves, hobbling away along the street.

Whepler did what he said he would do. He made some inquiries. He poked around. He really did try. But he didn't find any trace of that little girl. Her mother came around each morning for the rest of that week, and then twice a week for a time, then once a week, then every month or so.

One day, probably a year later, Whepler took a taxi up to the *barrio* where Señor Nevares lived. She was still there. She seemed to be doing okay. He tried to give her back her photo, the one of her daughter, but she wouldn't have it. She started to cry. She probably thought it would be bad luck, an admission that there really was no hope. So Whepler put it in his wallet and he has kept it there.

Poor old Jean-Marc. He never did find any Salvadorans to attend that dinner of his for the French ambassador. What he did instead was, with about an hour left to go before the dinner itself, he came barging into the lobby of the Camino Real, high as a kite, and he invited the entire foreign press corps to come on over to the room he'd rented at the Sheraton Hotel, just in order to fill the seats. And they all did.

Decked out in their jeans and hiking boots and polo shirts, they slogged over there aboard a fleet of taxis and rented cars. They filled the seats and wolfed down the free paella they were served and the wine. Then they took turns climbing to their feet and proposing long toasts in Spanish. *¡Viva Mitterrand! ¡Viva D'Aubuisson!* Long live French-Salvadoran solidarity! One toast after another, each more ridiculous and pompous than the one before....

Grant Lovsted went back to the States not long after that. He got married to some woman up there, and they have a couple of kids now, both girls. He was on the education beat at his paper, last Whepler heard. Most of the other hacks have moved on, too, replaced by newer faces, younger scribes. Whepler has stayed down here, though—he doesn't know why.

He used to get into a taxi sometimes, usually in the afternoon when it was sunny out and there was nothing else to do, when the city was glazed in that Salvadoran light, that polished, waxy glow. He'd go for a tour around town, tell the driver to slow down each time they came upon a group of children. He was looking for some echo, some semblance, of that girl. Maybe the police had let her go and she was being raised by strangers. Maybe she'd lost her memory—you never knew.

Then he realized she wouldn't be a child any more. She'd be a young woman, hard to recognize now. But still he finds himself scanning the faces in a crowd, wondering if he might see someone who looks like her, a girl miraculously returned to life. He never has seen her, never has caught a glimpse of her that he knows of. Yet he keeps his eyes peeled for that moment. And sometimes he still drops by different government offices, maybe once a month, asks questions about this girl named Alma Nevares. He doesn't expect any answers, just wants them to remember that someone wants to know.

He still thinks about heading home, or even putting in for a new assignment someplace else—possibly Delhi, maybe Cairo—but somehow he never manages to make up his mind. Something always gets in his way. There's seventy thousand dead civilians in this war so far, and the number just keeps on growing. The war never quits. They say it's going to end someday, but they don't say when.

Protection of World's Young:

Clocking out on Child Labor

~

BY

**AMELIA A.
NEWCOMB**

On the roadsides of Bombay or down crowded alleys in Manila or in the back rooms of Bogotá, children weave rugs, pound metal, or handle toxic substances, often for 14 hours a day and for little pay.

Their wares often end up in the world's largest importer—the United States—with American consumers unaware of their origin.

The global problem of child labor is often overlooked by the major powers. Now, however, there are signs of change.

The U.S., for the first time, has begun efforts to curb child labor overseas. The Senate has taken up a measure that would bar imports of child-made goods. Exporters in such places as India, Honduras, and Bangladesh are scrambling to change their labor practices.

The impact has been felt worldwide:

- In India, more than 130 carpetmakers are part of a "Rugmark" campaign that tags products as made without child labor.
- In Bangladesh, the Garment Manufacturers and Export Association formally agreed in November 1994 to ban hiring children under 14. They also agreed to make provisions for educating children who work full time.
- In the U.S., the Liz Claiborne company moved swiftly to inspect its manufacturing facilities after a Senate hearing in September 1994 revealed that child workers were involved in making the company's sweaters. Other companies have also been working on "codes of conduct" to address child labor.

"The U.S. is the world's largest importer of manufactured products from developing countries, and that puts a special onus on us to think of what we can do," says Robert Senser, who has written extensively on human rights in employment.

Congress failed to pass the Child Labor Deterrence Act in 1994, but advocates are optimistic for passage in the next session. In the meantime, recent events have focused new attention on how U.S. imports are manufactured overseas.

A U.S. Department of Labor report issued in September 1994 targeted 19 countries where at least 46 million children allegedly toil, making goods for the U.S. market. A follow-up report is expected next year to identify companies that produce or sell such items.

"This report has more impact because it comes from the U.S. government," says Sonia Rosen, who directed the study for the department's Bureau of International Labor Affairs. "It keeps the issue up on the radar screen."

The Labor Department report reveals an unsavory picture of how children contribute to their nations' exports.

In Colombia, about 800 000 children aged 12 through 17 are exposed to toxic substances while they process and harvest flowers for export. In South and Southeast Asia, where the International Labor Organization estimates at least half of all child workers live, children sew clothing during 14-hour days in crowded garment factories and knot carpets for hours on end in dusty huts.

In Africa, there is growing concern that more and more children will end up sewing garments or mining minerals as nations build factories and families flock to cities.

While rural child labor has long existed, urban child labor increased in the past two decades. The garment industry in Bangladesh, for example, among the largest suppliers of cotton clothing in the U.S., sprang to life about 10 years ago.

"In the past decade, the growth of children working has tracked closely the growth in exports," says Pharis Harvey, executive director of the International Labor Rights Education and Research Fund. A strong political and social tolerance for child labor servitude in many countries, he says, undermines efforts at reform.

Growing awareness of the connection between the push to export and the employment of minors has resulted in renewed pressure to enforce existing child-labor laws in the offending countries. Numerous nongovernmental organizations, such as the South Asian Coalition on Child Servitude, based in India, have long pressed for improved access to education for poor children, who often end up working as bonded laborers, and have stepped up campaigns to get children out of factories. The ILO launched its first major campaign against child labor in 1992, channeling funds for educational and technical assistance to eight countries.

But when the Child Labor Deterrence Act was introduced, also in 1992, a number of countries sat up and took notice. And this, say many child-rights advocates, is proof positive that the U.S. can influence the use of child labor overseas.

Critics of attempts to ban child-made products decry such measures as protectionist and ignorant of the reality that families in many developing countries rely on a child's income. But supporters counter that many of the industries targeted no longer exist in the U.S.

Also, Bill Goold, legislative director for Rep. George E. Brown Jr. (D) of California, sponsor of the House version of the bill, argues that, rather than hurting children or families who desperately need the income, a ban could assist countries in breaking long-standing cycles of poverty.

"When kids keep working, adults don't get jobs," says Mr. Goold. In the rug sector, he points out, there is acute adult unemployment. "It's a false premise to say kids have to work or the family starves."

But getting the message across isn't easy. Sen. Tom Harkin (D) of Iowa, who introduced the bill, was pilloried in Bangladesh as well as in India, with those governments and some nongovernmental organizations lambasting the bill as imperialistic.

Nevertheless, in anticipation of foreign visits at the time, some of the larger Bangladeshi garment factories moved to comply with existing but unenforced child-labor regulations. And in addition to agreeing not to hire any more child laborers under the age of 14, the Bangladesh Garment Manufacturers and Export Association promised that children aged 12 to 14 will move to part-time work and part-time school, while children under 12 will be sent to school full time.

Ideally, activists say, consumer pressure and legislative

action combined will end reliance on child labor in certain export growth industries. They are also counting on many American consumers becoming informed about products made by children.

"I find there are a lot of people who don't know what's going on, and when I do talk with them, the reaction is sympathetic," Mr. Senser says. "President Clinton just ordered a ban on imports involving rhino horn and tiger bones. What's wrong with putting something in about children?"

The Children of Bogota

BY

PATRICK

LANE

The first thing to understand, Manuel says,
is that they're not children. Don't start feeling
sorry for them. There are five thousand
roaming the streets of this city

and just because they look innocent
doesn't make them human. Any one
would kill you for the price of a meal.
Children? See those two in the gutter

behind that stall? I saw them put out
the eyes of a dog with thorns because
it barked at them. Tomorrow it could be you.
No one knows where they come from

but you can be sure they're not going.
In five years they'll be men and tired of killing
dogs. And when that happens you'll be the first
to cheer when the carabineros shoot them down.

Without Hands

~

BY

LORNA

CROZIER

*(In memory of Victor Jara, the Chilean musician
whose hands were smashed by the military to
stop him from playing his guitar and singing for
his fellow prisoners in the Santiago stadium.
Along with thousands of others, he was tortured
and finally killed there in September, 1973.)*

All the machines in the world
stop. The textile machines, the paper machines,
the machines in the mines turning stones to fire.
Without hands to touch them, spoons, forks and knives
forget their names and uses, the baby is not bathed,
bread rises on the stove, overflows the bowl.
Without hands, the looms
stop. The music
                  stops.
The plums turn sweet and sticky and gather flies.

Without hands
            without those beautiful conjunctions
those translators of skin, bone, hair
two eyes go blind
two pale hounds sniffing ahead and doubling back
to tell us
        of hot and cold or the silk of roses after rain
are lost
        two terns feeling the air in every feather
are shot down.

Without hands my father doesn't plant potatoes
row on row, build a house for wrens,
or carry me
from the car to bed
when I pretend I'm sleeping.
On wash-days my mother doesn't hang clothes
on the line, she doesn't turn the pages of a book
and read out loud,
or teach me how to lace my shoes.

Without hands my small grandmother
doesn't pluck the chicken for our Sunday meal
or every evening, before she goes to sleep,
brush and brush her long white hair.

Our Father Who Art in Heaven

BY

JOSÉ LEANDRO

URBINA

While the sergeant was interrogating his mother and sister, the captain took the child by the hand to the other room.

—Where is your father? he asked.

—He's in heaven, whispered the boy.

—What's that? Is he dead? asked the captain, surprised.

—No, said the child. Every night he comes down from heaven to eat with us.

The captain raised his eyes and discovered the little door in the ceiling.

The Purple Children

BY

EDITH

PARGETER

~

he outrage took place at eleven o'clock on a moonless night, before the stars began to silver the white walls of the church. It was the sentry at the rear gate of the Town Hall, an eighteen-year-old new to the town, who was singled out as the weakest spot in the defences. Half-dozing on his walk back and forth across the gate, he heard the most innocent sound in the world, a girl's voice calling softly: "Puss, puss, puss..."

As he started into wakefulness with the exaggerated attention which made the walls seem higher and the night darker, a little figure with the light running steps of a child darted towards the gate, and halted with her hands locked upon the bars. He saw how slight she was, and how young, no more than fifteen. Her frock was dark, probably black like so many of them here. She turned on him a face which was only a silvery oval and a dark shining of eyes, and he thought he saw about it the shadowy movement of unkempt locks darker than the darkness.

"You can't go in there," said the sentry gruffly. "You ought to be in the house at this time of night. Don't you know there's a curfew?"

"I *was* in the house. I only came out because of my cat; she got out when I went to bring in wood, and I couldn't catch her. She's young, she runs away. It's no use telling *her* there's a curfew."

"She'll come back in the morning," said the sentry awkwardly. "They always do. You go home like a good girl, and don't you risk running about here in the dark. Somebody might think you were up to something."

"But she might not come back. She's never been out at night before. I could get her now, if you'll let me. She ran through there into the courtyard. Won't you please help me to catch her?"

The boy felt the small, cold hand laid entreatingly upon his arm. She was only a kid, she hardly came up to his shoulder, and she was beginning to sniff ominously. He couldn't see any harm in it. He had orders to treat the natives politely and considerately, as long as they weren't making trouble, and what trouble could this waif possibly make?

"I can't! I should get into trouble if anybody found out..."

"Well, who's going to find out? All I want is to get my cat. You'll be there close to me every minute, you can see every move I make. And you've got a gun—I don't see what you have to be afraid of. Oh, do please help me!"

He hadn't meant to do it, but somehow he had set his hand to the bars beside hers, and thrust the gate open before her. "Well, be quiet about it, can't you, or somebody'll hear us. Come on, quick, and get her, and get yourself out of here."

She slipped past him like a shadow. He turned his back on the gate and the silent, dark lane outside and pressed at her shoulder as she flitted into the darkest corner of the yard, where the outhouses leaned together in a huddle of shadows, and the steps plunged down to the cellar. Behind them the tall bulk of the Town Hall shut off the awaking stars, and the ropes of the flagstaff creaked faintly in the wind which never stilled in the upper air.

"There she is!" whispered the girl triumphantly, and darted forward and was lost among the deeper shadows under the wall. And there really was a cat, the sentry saw with relief and satisfaction, a thin little tabby skipping from darkness to darkness, evading them with the light, unhurried insolence of cats everywhere. It took them ten minutes to run it to earth at the foot of the cellar steps, against the closed door. The girl snatched it up and held it struggling in her arms, and looked up at him under the black tangle of her hair with a wild smile.

"Thank you! Now I'll go home. You were very kind to let me come in!"

But she did not move; she stood looking at him still, her eyes enormous and shy and wary. When she looked at him like that he felt how alien he was in this place, and even her thanks could not compensate him for the quiet, patient hatred of her people. She let her body touch his, her sharp little shoulder leaning for a moment into the hollow of his arm, which moved of itself to hold her.

Then they both heard, clear through the silence, the sudden light impact of feet, as though someone had dropped from the high wall.

The sentry spun round and went up the cellar steps three at a time, just in time to see the figure of a boy disentangle itself from the severed ropes of the flagstaff, and run head-down for the gate.

The shout of rage and alarm was out of him before he knew it, and after that there was no hope of keeping it all quiet and pretending that no one had got past him during the night; the only chance he had was to get at least one prisoner to show for it. He hurtled after the racing boy, hauling the loaded spray-gun round from his shoulder in flight to bring it to bear upon the fugitive. He heard the girl scream, and was startled because the sound came from only a yard or two behind him, where silently, wildly, she was running, too. When she saw him check for an instant to steady the gun, she ran past him and flung the cat sprawling and clawing in his face.

He threw up his left arm to cover his eyes, and swerved aside, firing the gun blindly. The spray spattered darkly over her cheek and her spread hands, but she had gained the few yards she needed for herself and her partner, and she flew through the gate and pulled it to with a clang. Before the sentry could fling off the cat and wipe his eyes clear of the blood from his scratched forehead, both the fugitives were snatched away into the silence and darkness of the little streets.

People came pouring into the courtyard from three doors. They found the sentry mopping his face, a long, violet stain upon the ground, and the coils of severed rope dangling at the foot of the flagstaff. They got the major out of bed, and the sentry reported to him with every excuse he could think of, though the sum of them all sounded thin enough.

"She was only a kid about fifteen. I didn't think she could be up to anything, sir. She was looking for her cat."

The major had been in the country for over a year, and was accustomed to the local style of warfare, to the ugly demands it made upon him, and the satisfaction he sometimes felt in their ugliness, which frightened and depressed him more than anything else. He stood gazing at the boy without rancour.

"They're always kids of fifteen. Haven't you learned that yet?"

"But there was a cat, sir, that was true, anyhow."

"That skinny tabby," said the major wearily, "belongs to the caretaker. I imagine its appearance was a stroke of luck. Or she may have seen it before she made up her story and began calling. Well, you seem to have spent practically a quarter of an hour being civil to her, I take it you can pick her out again?"

The sentry was too frightened of his own side, by this time, to retain much resentment against the enemy; his fear even drew him into a kind of distant alliance with them. He said: "No, sir, I don't think I could. It was pretty dark there under the wall. There's scores of them that same build, thin as a monkey."

"And scores of them with purple hands and faces? At least you had the sense to fire your charge. That ought to give her one distinguishing feature, don't you think?"

The sentry looked at the long dark stain like blood upon the stones, and was filled with a treasonable but unmistakable regret. "I'm sorry, sir," he lied. "It was just then she threw the cat, it put me off proper. I reckon I missed her."

"Then why," asked the major gently, "did she drip violet dye practically all the way to the gate? He marked the last infinitesimal spot in the light of his torch. "A heavier charge, and we might have been able to follow her all the way home. Did you mark the boy, too?"

"No, sir. He was well out of range, only he turned back to catch hold of her hand." It was the first time he had fully realized all that he had seen. Regret rose in him like a tidal sea. "They haven't done anything all that bad, sir—it's only a flag!"

The major smiled. When this boy was forty instead of eighteen he would no longer make the absurd mistake of speaking of "only" a flag. "Whoever it was, he's left about ten feet of the flagstaff coiled round with barbed-wire as he came down. You must have been very absorbed in your cat-hunt. And he must have spent a long time practising the movements involved, before he could reproduce them at that speed. Yes, I should like to congratulate that boy! But when we've found her we shall have found him, too. We'll try the grammar school first," he said, smiling to himself, beginning to feel the terrifying satisfaction of hate reacting against hate. "If she isn't there, we'll look up the girls who don't answer the register. We shan't have to look any farther."

In the shed behind Pablito's father's shop Mariposa knelt over a pan of water, scouring with a handful of wet sand at the backs of her hands. The water lay in her palms as she rinsed them as clear as it had come from the well. Juanito held the torch close, keeping his body between its light and the covered window. Teo crouched on his heels, his head bent close to Mariposa's, his cheek brushed occasionally by her swinging hair.

"It's no use," she said, letting her hands lie quiet in the wet skirt of her dress and looking up at him with enormous black eyes. The misshapen blotches of purple ate away half her face into shadow. Behind her all the silent, intent partisans drew closer with a long sigh. "It won't come out," she said with the calm of despair. "Now they have only to look for me—I can't be hidden. Teo, what am I to do?"

"If they find you," he said, taking her stained hands in his, "they find me, too."

"That's foolish! You'll be needed again. And besides, they'd beat you; they'll only imprison me. No, it was great luck that you were not splashed like me—don't be so ungrateful as to throw it away." But she was very frightened. He felt the small, wet hands, hot with scouring, tremble in his own.

"I will not let you bear it alone! We were all in this thing together. When we two drew the lots we drew the danger with them, as well as the privilege."

"They'll come straight to the school," said Juanito. "Perhaps if you stay at home and take care not to be seen..."

"For how long? said Teo shortly. "You see the marks will have to wear away gradually. Do you think she can be hidden for months?

"But they may give up in a week or two. She need not be hidden from our own people, only from *them*."

"If they do not find me in the school," said Mariposa with authority, "they will want to see the register and find out who is missing. It is only another way of being set apart. I think I would rather be there to face them. It is not I who will have cause to be ashamed." But her body shook and her hands contracted in the boy's hands, because she knew she would still be afraid.

"If we tried linen-bleach," said Luz timidly, "do you think it would remove it?"

Esperanza shook her head. "It's an old vegetable dye, nothing will fetch it out, that's why all the dyers still use it. My father is a dyer," she said sadly. "I know!"

Teo stood up slowly, still holding the thin, marred hands in his. All the intent and anxious eyes settled upon him and clung hopefully, because he had begun to smile. He looked down and smiled into Mariposa's eyes, and his thin, brown face relaxed into a reassuring tranquility. "Come!" he said, drawing her up by the hands." I have an idea! Come, all of you! It will be hours yet to daylight, we have a little time. Don't be afraid, they won't find you! They'll never find you!"

In the morning light the major looked out his window and saw the silvery coils of new barbed-wire like a guardian serpent about the flagstaff, and above, afloat upon the restless wind, the expected flag, an enemy that could never be silenced. It would soon be down, of course. It could not be nailed to the staff, there had not been time, and silence had been essential. Yes, it would soon be down; the only trouble was that it would go up again somewhere else. It always did.

He had spent a year of his life searching the little houses of these towns for explosives and arms, for subversive literature in the native tongue, for wanted men on the run; but on every occasion the circumstances of the search became a little meaner, and more humiliating. Now it was a little girl with a face stained by vegetable dye, who had made a fool of a homesick boy and helped another boy to raise once again the ubiquitous flag. The major felt an impatience to have the miserable business finished; but by daylight he no longer mistook for anger and hatred what was, after all, nothing but disgust and exasperation.

The grammar school opened at eight o'clock; at half past eight the major presented himself there with a sergeant and two men.

He was punctilious in waiting in the entrance hall while the headmaster was fetched out to him.

"I need hardly tell you why I'm here," he said. "No doubt you've already seen the flag over the Town Hall. We intend to make an example this time. If you allow your children to move up into the front line, you must consider that it is you yourselves who inflict their punishments upon them. We would infinitely rather deal with you."

"We would infinitely rather that you did," agreed the headmaster, his spectacles a little askew on his antique and aquiline nose. "You must do what you feel to be your duty. But so must our children. Would you like to begin with the little ones? Forgive me, but your gambit leads me to suppose that you are looking for someone more than usually embarrassing as an opponent."

The major would have liked to think of a cutting reply, but the situation had placed irony clean out of his reach. "I am looking for a girl of about fifteen. You may not know that we have recently adopted the use of a spray-gun loaded with one of the local vegetable dyes. The girl will be stained purple. This time I can promise you there will be no collective punishments. This time it will not be necessary."

"Purple!" said the schoolmaster reflectively. "A royal color. Also the color of mourning. A nice choice!"

"There was also a boy, who will not be so immediately recognizable. But I think it hardly matters. Once we have our hands on the girl he will come forward of his own will."

"I see you have not entirely wasted your time with us," said the schoolmaster politely. "Very well! You wish to inspect our senior grades? I have kept them assembled in the hall for you. Please!"

The major strode across the polished lobby, the sergeant and his men keeping step behind him. The headmaster, advancing his hand to open the door, leveled one sudden, glittering glance into the eyes of the invader, and it seemed for an instant that what he felt for him was no longer simple antagonism, but almost pity. Then he pushed the door wide, and stood back for his visitors to enter.

The major marched over the threshold with the briskness of complete confidence, almost of triumph. Fifty-three young heads, with marvelous unanimity, were raised to confront him, the challenging light of fifty-three pairs of dark, wide, Byzantine eyes bristled at him like bayonets, and he checked in his stride and wrenched himself sidewise into stillness, as though he had indeed run his beribboned breast into a thicket of steel. He had come looking for a marked outcast. He beheld a regiment, a Pyrrhic phalanx of embattled children, all their delicate olive faces spattered from forehead to chin with the resplendent purple of royalty and mourning.

Just Lather, That's All

BY

HERNANDO

TÉLLEZ

Translated by

Donald A. Yates

He said nothing when he entered. I was passing the best of my razors back and forth on a strop. When I recognized him I started to tremble. But he didn't notice. Hoping to conceal my emotion, I continued sharpening the razor. I tested it on the meat of my thumb, and then held it up to the light. At that moment he took off the bullet-studded belt that his gun holster dangled from. He hung it up on a wall hook and placed his military cap over it. Then he turned to me, loosening the knot of his tie, and said, "It's hot as hell. Give me a shave." He sat in the chair.

I estimated he had a four-day beard. The four days taken up by the latest expedition in search of our troops. His face seemed reddened, burned by the sun. Carefully, I began to prepare the soap. I cut off a few slices, dropped them into the cup, mixed in a bit of warm water, and began to stir with the brush. Immediately the foam began to rise. "The other boys in the group should have this much beard, too." I continued stirring the lather.

"But we did all right, you know. We got the main ones. We brought back some dead, and we've got some others still alive. But pretty soon they'll all be dead."

"How many did you catch?" I asked.

"Fourteen. We had to go pretty deep into the woods to find them.

But we'll get even. Not one of them comes out of this alive, not one."

He leaned back on the chair when he saw me with the lather-covered brush in my hand. I still had to put the sheet on him. No doubt about it, I was upset. I took a sheet out of a drawer and knotted it around my customer's neck. He wouldn't stop talking. He probably thought I was in sympathy with his party.

"The town must have learned a lesson from what we did the other day," he said.

"Yes," I replied, securing the knot at the base of his dark, sweaty neck.

"That was a fine show, eh?"

"Very good," I answered, turning back for the brush. The man closed his eyes with a gesture of fatigue and sat waiting for the cool caress of the soap. I had never had him so close to me. The day he ordered the whole town to file into the patio of the school to see the four rebels hanging there, I came face to face with him for an instant. But the sight of the mutilated bodies kept me from noticing the face of the man who had directed it all, the face I was now about to take into my hands. It was not an unpleasant face, certainly. And the beard, which made him seem a bit older than he was, didn't suit him badly at all. His name was Torres. Captain Torres. A man of imagination, because who else would have thought of hanging the naked rebels and then holding target practice on certain parts of their bodies? I began to apply the first layer of soap. With his eyes closed, he continued. "Without any effort I could go straight to sleep," he said, "but there's plenty to do this afternoon." I stopped the lathering and asked with a feigned lack of interest: "A firing squad?" "Something like that, but a little slower." I got on with the job of lathering his beard. My hands started trembling again. The man could not possibly realize it, and this was in my favor. But I would have preferred that he hadn't come. It was likely that many of our faction had seen him enter. And an enemy under one's roof imposes certain conditions. I would be obliged to shave that beard like any other one, carefully, gently, like that of any customer, taking pains to see that no single pore emitted a drop of blood. Being careful to see that the little tufts of hair did not lead the blade astray. Seeing that his skin ended up clean, soft, and healthy, so that passing the back of my hand over it I couldn't feel a hair. Yes, I was secretly a rebel, but I was also a conscientious barber, and proud of the preciseness of my profession. And this four-days' growth of beard was a fitting challenge.

I took the razor, opened up the two protective arms, exposed the blade and began the job, from one of the sideburns downward. The razor responded beautifully. His beard was inflexible and hard, not too long, but thick. Bit by bit the skin emerged. The razor rasped along, making its customary sound as fluffs of lather mixed with bits of hair gathered along the blade. I paused a moment to clean it, then took up the strop again to sharpen the razor, because I'm a barber who does things properly. The man, who had kept his eyes closed, opened them now, removed one of his hands from under the sheet, felt the spot on his face where the soap had been cleared off, and said, "Come to the school today at six o'clock." "The same thing as the other day?" I asked horrified. "It could be better," he replied. "What do you plan to do?" "I don't know yet. But we'll amuse ourselves." Once more he leaned back and closed his eyes. I approached him with the razor poised. "Do you plan to punish them all?" I ventured timidly. "All." The soap was drying on his face. I had to hurry. In the mirror I looked down the street. It was the same as ever: the grocery store with two or three customers in it. Then I glanced at the clock: two-twenty in the afternoon. The razor continued on its downward stroke. Now from the other sideburn down. A thick, blue beard. He should have let it grow like some poets or priests do. It would suit him well. A lot of people wouldn't recognize him. Much to his benefit, I thought, as I attempted to cover the neck area smoothly. There, for sure, the razor had to be handled masterfully, since the hair, although softer, grew into little swirls. A curly beard. One of the tiny pores could be opened up and issue forth its pearl of blood. A good barber such as I prides himself on never allowing this to happen to a client. And this was a first-class client. How many of us had he ordered shot? How many of us had he ordered mutilated? It was better not to think about it. Torres did not know that I was the enemy. He did not know it nor did the rest. It was a secret shared by very few, precisely so that I could inform the revolutionaries of what Torres was doing in the town and of what he was planning each time he undertook a rebel-hunting excursion. So it was going to be very difficult to explain that I had him right in my hands and let him go peacefully—alive and shaved.

The beard was now almost completely gone. He seemed younger, less burdened by years than when he had arrived. I suppose this always happens with men who visit barber shops. Under the stroke of my razor Torres was being rejuvenated—rejuvenated because I am a good barber, the best in the town, if I may say so. A little more lather here, under his chin, on his Adam's apple, on this big vein. How hot it is getting! Torres must be sweating as much as I. But he is not afraid. He is a calm man, who is not even thinking about what he is going to do with the prisoners this afternoon. On the other hand I, with this razor in my hands, stroking and restroking this skin, trying to keep blood from oozing from these pores, can't even think clearly. Damn him for coming, because I'm a revolutionary and not a murderer. And how easy it would be to kill him. And he deserves it. Does he? No! What the devil! No one deserves to have someone else make the sacrifice of becoming a murderer. What do you gain by it? Nothing. Others come along and still others, and the first ones kill the second ones and they the next ones and it goes on like this until everything is a sea of blood. I could cut this throat just so, zip! zip! I wouldn't give him time to complain and since he has his eyes closed he wouldn't see the glistening knife blade or my glistening eyes. But I'm trembling like a real murderer. Out of his neck a gush of blood would spout onto the sheet, on the chair, on my hands, on the floor. I would have to close the door. And the blood would keep inching along the floor, warm, ineradicable, uncontainable, until it reached the street, like a little scarlet stream. I'm sure that one solid stroke, one deep incision, would prevent any pain. He wouldn't suffer. But what would I do with the body? Where would I hide it? I would have to flee, leaving all I have behind, and take refuge far away, far, far away. But they would follow until they found me. "Captain Torres' murderer. He slit his throat while he was shaving him—a coward." And then on the other side. "The avenger of us all. A name to remember. (And here they would mention my name.) He was the town barber. No one knew he was defending our cause."

And what of all this? Murderer or hero? My destiny depends on the edge of this blade. I can turn my hand a bit more, press a little harder on the razor, and sink it in. The skin would give way like silk, like rubber, like the strop. There is nothing more tender than human skin and the blood is always there, ready to pour forth. A blade like this doesn't fail. It is my best. But I don't want to be a murderer, no sir. You came to me for a shave. And I perform my work honorably.... I don't want blood on my hands. Just lather, that's all. You are an executioner and I am only a barber. Each person has his own place in the scheme of things. That's right. His own place.

JUST LATHER, THAT'S ALL : **183**

Now his chin had been stroked clean and smooth. The man sat up and looked into the mirror. He rubbed his hands over his skin and felt it fresh, like new.

"Thanks," he said. He went to the hanger for his belt, pistol and cap. I must have been very pale; my shirt felt soaked. Torres finished adjusting the buckle, straightened his pistol in the holster and after automatically smoothing down his hair, he put on the cap. From his pants pocket he took out several coins to pay me for my services. And he began to head toward the door. In the doorway he paused for a moment, and turning to me he said:

"They told me that you'd kill me. I came to find out. But killing isn't easy. You can take my word for it." And he headed on down the street.

Third Wave War

BY
ALVIN AND HEIDI TOFFLER

~

Long before Leonardo da Vinci began toying with the idea of flying machines and fantastic forerunners of the tank, the rocket, and the flamethrower, creative minds conjured up weapons of the future.

Today, despite cutbacks in military spending in many (though by no means all) countries, military imagination is still hard at work. If we ask thoughtful military men what their forces will need in the years ahead, they pull out of their desk drawer a dazzling list of dream weapons. Few of these will ever actually come into being. But some of them *will* materialize and play their part in Third Wave warfare.

What most nations now want are smarter weapons, beginning with sensors. American military planners hunger for next-generation sensors able to detect fixed and moving objects from 500 to 1000 miles [805 to 1610 km] away. Such sensors would be mounted on aircraft, drones, or space vehicles, but more important, they would be under the decentralized control of theatre commanders, who would be able to move them around as needed and customize the information streaming in from them. This smart sensor of the near-term future would bring together or "fuse" different kinds of fine-grained data, synthesize it, and check it against many kinds of data bases. The result would be better early warning, more refined targeting, and improved damage assessments. Sensors are top priority.

On the ground, the army wishes to replace stupid, inert mines with smart mines that don't wait for an enemy tank to roll over them. Instead, the "dream mine" would acoustically scan the area around it, compare engine sounds and earth rumbles against a list of vehicle types, identify the target, use an infrared sensor to locate it, and then fire a shaped charge at it.

The U.S. Army is also looking into "smart armor" for its own tanks. As an incoming projectile approaches, a mesh of sensors mounted outside the skin of the tank would measure and identify the type of round and instantly communicate that information to an on-board computer. Tiny explosive "tiles" on the outside of the tank would be fired off by the computer to deflect or destroy the inbound shell. Such advanced armor would fend off either kinetic or chemical warheads.

Other planners picture an all-electric battlefield, spelling the end of the Age of Gunpowder for artillery. In this scenario, electricity propels the shell and electronics guides it to its target. All vehicles are electrical, recharged, perhaps, by aircraft that fly over them and zap energy to them.

The individual soldier is also reconceptualized. According to Maj. Gen. Jerry Harrison, former head of research and development laboratories for the U.S. Army, the soldier should no longer be viewed "as something you hang a rifle on, or that you hang a radio on, but as a system."

Already under research is the concept of SIPS—the Soldier Integrated Protective Suit. This is a "suit" that would offer protection against nuclear, chemical, or biological weapons, provide the soldier with night-vision goggles and a heads-up display. It would also include an aiming system that tracks eye movements so that it can automatically point the gun at whatever the soldier is looking at.

These and additional capabilities would all be integrated into a suit that is right out of a Hollywood special effects department—an intelligent exo-skeletal suit that learns to perform the soldier's repetitive tasks so he or she can march 10 miles [16 km] and doze off while doing it...a suit that amplifies the strength of the wearer several-fold. As General Harrison puts it, "I want to put this guy in some sort of exoskeletal suit that will allow him to leap tall buildings with a single bound." The allusion to Superman is clear.

The soldier inside this smart suit, however, is not an over-muscled, small-brained cartoon character but an intelligent man or woman capable of processing huge amounts of information, analyzing it, and taking resourceful action based on it.

This vision of every soldier a Superman or Schwarzenegger, or more accurately a Terminator, is taken seriously enough for a group of researchers to have formed around the concept at the U.S. Army's Human Engineering Laboratory in Aberdeen, Maryland.

According to Maj.-Gen. William Forster, director of combat requirements in the Pentagon, the ultimate object of the work on SIPS is "to increase the effectiveness of the individual so that you need fewer soldiers. The fewer 'soft-skin' soldiers we have out there, the fewer the casualties."

Science fiction-like or not, Forster notes, "The Exo-Skeleton or Exo-Man, even though it is far-out, all these things are within the known laws of physics. You don't have to change the laws to do them. The real trick is doing them economically and reliably."

Also within the framework of known laws are even more remarkable possibilities. Micro-machines, for example. Today the first micro-machines are just being patented—for example, an electric motor less than a millimetre long that could, according to Prof. Johannes G. Smits, drive a robot the size of an ant.

"Imagine what you could do with an ant if you could control it," says Smits, an electrical engineer at Boston University, who holds the patent on the new motor. "You could make it walk into CIA headquarters." The energy to drive the micro-robot could come from a micro-microphone that converts sound into energy.

It doesn't require much imagination to appreciate what an infestation of robotic ants could do to any enemy's radar installation, or to aircraft engines or to a computer centre.

Such micro-machines, however, are huge, hulking giants compared with the nano-machines to come. If micro-machines are small enough to manipulate individual cells, nano-machines can manipulate the molecules of which cells are built. Nano-robots would be small enough to operate like submarines in the bloodstream of humans, and presumably could, among other things, be used to perform surgery at the molecular level.

Work on nano-technology is under way in the United States and Japan, where researchers Yotaro Hatamura and Hiroshi Miroshita have prepared a study on Direct Coupling Between Nanometer World and Human World. According to a survey of 25 scientists working on nanotech, within the next 10 to 25 years we will not merely be able to create devices at the molecular scale, but will be able to make them self-replicating—meaning we can breed them.

Here we approach the "self-reproducing war machines" alluded to earlier. For example, the smart sensors we have been talking about so far are near-term extensions of current technology. But a generation from now, says a physicist at the RAND Corporation, "we start looking at sensors that...can burrow into communications systems, or sensors that can lie there for 20 years, just ticking away, ready to be remotely activated. They could be the size of a brilliant pinpoint under the ground."

Imagine then, super-smart sensors and mines, a few nanometers in size, that can, as suggested in the preceding paragraphs, reproduce themselves. Now picture a scenario in which a global police force seeds them over a pariah state and programs them to replicate to a given density in militarily sensitive regions. Virtually undetectable and harmless, the mines could be armed selectively from the outside by tiny pulses of energy. At which point the local Saddam Hussein is told to close down his chemical weapons plant or see all his military bases erupt. Unless, of course, the enemy reprograms them. Or they refuse to stop breeding. Of course, all this is, at this point, just fantasy. But so were Leonardo's flying machines when he drew them.

We need not wait for self-breeding nano-technology, however, to face novel terrors. Long before then the diffusion of swiftly advancing scientific knowledge threatens to turn conventional chemical and biological weaponry into the so-called "poor-man's nuclear bomb."

While it still remains cumbersome to handle and deliver most chemical or biological weapons without endangering one's own forces, that is hardly likely to inhibit the Pol Pots or Saddam Husseins of tomorrow. The world has justifiably begun to worry about chemical and biological weapons programs in countries like Libya, India, Pakistan, China, and North Korea, not to mention Iraq, many of which may face political and economic instability in the decades to come.

In January, 1993, with much self-congratulation, after a quarter century of negotiation, 120 nations met in Paris to sign the Chemical Weapons Convention. Theoretically it bans the production and storage of chemical arms. A watching body, the Organization for the Prohibition of Chemical Weapons (OPCW), was established to police the agreement. Its inspectors will have greater powers than those enjoyed by the International Atomic Energy Agency (IAEA) until now. But 21 members of the Arab League refused to join in the agreement until Israel did. Iraq sent no one to the meeting. And the Convention actually does not come into effect until half a year after fully 65 nations ratify it.

As for biological warfare agents—in many ways the worst of the weapons of mass destruction—it is now known that work on offensive bio-war weapons continued in the Soviet Union long after it signed a 1972 treaty outlawing such arms; long after these activities were denied by Gorbachev; long after the Soviet state collapsed and was replaced by Russia; and even after Yeltsin publicly ordered germ warfare research ended. That work included—and may still include— a search for a genetically engineered "super-plague" that could wipe out half the population of a small city in short order.

Who, in a country torn apart politically and on the edge of anarchy, controls the pathogens that still, no doubt, remain in the laboratories of the former Soviet Union? And how safe are they?

In 1976, the Soviets, undoubtedly aware of the horrors breeding in their own laboratories, called for international bans on exotic arms. They warned, at that time, of the hideous possibility of race-specific weapons—genetically engineered to single out and decimate only the members of selected ethnic groups—the ultimate genocidal weapon for race war. In 1992 Bo Rybeck, director of the Swedish National Defense Research Institute, pointed out that as we become able to identify the DNA variations of different racial and ethnic groups, "we will be able to determine the differences between blacks and whites and orientals and Jews and Swedes and Finns and develop an agent that will kill only (a particular) group. One can imagine the uses to which such technology might be put by "ethnic cleansers" of tomorrow.

The warning about race-specific weaponry takes on new urgency in light of recent scientific advances connected with the Human Genome Initiative, which aims at unlocking the secrets of DNA. Taken a step further, it conjures up the use of bioengineering or genetic engineering to alter soldiers or to breed "para-humans" to do the fighting.

And then there is ecological warfare. When Saddam Hussein torched the Kuwaiti oil fields, he was only doing what the Romans did when, according to some, they salted the fields of Carthage, and what the Russians did to their own fields during World War II when they pursued their "scorched earth" policy to deny food to the Nazi invaders. And, indeed, what the United States did with the use of defoliants in Vietnam.

These acts are primitive compared with some of the imaginable (and imagined) possibilities of sophisticated ecological weaponry. For example, triggering earthquakes or volcano eruptions at a distance by generating certain electromagnetic waves; deflecting wind currents; sending in a vector of genetically altered insects to devastate a selected crop; using lasers to cut a custom-tailored hole in the ozone layer over an adversary's land; and even modifying weather.

Lester Brown, of the Worldwatch Institute, a leading environmental think tank in Washington, D.C., pointed out as far back as 1977 that "deliberate attempts to alter the climate are becoming increasingly common," raising the prospect of "meteorological warfare as countries that are hard-pressed to expand food supplies begin to compete for available rainfall."

Evan's Progress

BY

PAUL

MCAULEY

O n Evan's eighth birthday, his aunt sent him the latest smash-hit biokit, *Splicing your own Semi-sentients*. The box-lid depicted an alien world throbbing with weird, amorphous life; a double helix spiralling out of a test-tube was embossed in one corner. "Don't let your father see that," his mother said, so Evan took it out to the old barn, set up the plastic culture trays and vials of chemicals and retroviruses on a dusty workbench in the shadow of the shrouded combine-harvester.

His father found Evan there two days later, watching the amoebae he'd created coalesce into a slimy blob soon to be transformed into a jellyfish. Evan's father dumped the culture trays and vials in the yard and made Evan pour a litre of industrial grade bleach over them. The acrid stench made Evan cry.

That summer, the leasing company foreclosed on the livestock that Evan's father rented. The rep who supervised repossession of the supercows drove off in a big car with the test-tube and double-helix logo on its door.

Next year the wheat failed, blighted by a particularly virulent disease. Evan's father couldn't afford the new resistant strain, and the farm went under.

~

Evan lived with his aunt in the capital. He was 15. He had a street bike, a plug-in computer, and a pet microsaur (also known as a triceratops with purple funfur). Buying the special porridge—all the microsaur could eat—took half of Evan's weekly allowance. So he let his best friend inject the pet with a bootleg virus to edit out its dietary dependence. It was only a partial success: the triceratops no longer needed its porridge but developed epilepsy triggered by sunlight. It shed fur in great swatches, so Evan abandoned it in a nearby park. Microsaurs were out of fashion anyway. There were dozens wandering the park. Soon they disappeared, starved to extinction.

The day before Evan graduated, his sponsor firm called to say that he wouldn't be doing research after all. There had been a change of policy: the previously covert gene-wars was going public. When Evan began protesting, the woman said sharply, "You're better off than many. With a degree in molecular genetics you'll make a sergeant at least."

The jungle below was a vivid green blanket in which rivers made silvery forked-lightnings. Evan leaned from the helicopter's hatch; the harness dug into his shoulders. He was 23, a technology sergeant. It was his second tour of duty.

Flashes on the ground. Evan hoped the peasants only had Kalashnikovs. Last week a sniper had downed a 'copter with an antiquated SAM: the tech sergeant had been too busy aiming sticky virus-suspension spray to kill the maize crop.

Afterwards, the pilot, an old-timer, said over the intercom, "Business gets dirtier every day. In the past we just used to take a leaf from Third World plants and clone it to get our supercrops. You couldn't really call it theft. But this stuff...I always thought war was bad for business."

"Not in this case," Evan said. "Our company owns the copyright to the maize that those peasants are growing. They haven't got licences to grow it."

"Man, you're a real company guy," the pilot said admiringly. "I bet you even know what country this is."

"Since when were countries important?"

Rice fields spread across the floodplain like a hand-stitched quilt. In every paddy, peasants bent over their reflections planting seedlings.

In the centre of the UNESCO delegation, the Minister for Agriculture stood under a black umbrella held by an aide, explaining why his country was starving to death after a record rice crop. Evan was at the back of the little crowd, bareheaded in the warm drizzle. He wore a smart one-piece suit, yellow overshoes. He was 28 and had spent two years infiltrating UNESCO for his company.

The Minister was saying, "We have to buy pesticide-resistant seed to compete with our neighbours, but the rice has to be exported to service our debt. We are starving in the midst of plenty."

Evan stifled a yawn. Later, at a reception he got the Minister on his own. The man was slightly drunk.

"Look in our cities," the Minister said slurring his words. "Every day a thousand refugees pour in from the countryside. We have kwashiokor, beri-beri."

Evan popped a canapé in his mouth. (One of his company's new lines, it squirmed lasciviously before he swallowed it.) "I may be able to help," he said. "The people I represent have a new yeast that fulfils all dietary requirements and will grow on a simple medium."

"How simple?"

Evan explained as the Minister, no longer drunk, steered him towards the terrace. The Minister said, "You understand this must be confidential. Under UNESCO rules…"

"There are ways around that. We already deal with five countries that have trade imbalances similar to your own. We lease the genetic material of the yeast to support governments who look favourably on our other products…"

~

The gene pirate was showing Evan his facility for editing genetic material, when the slow poison finally hit him. They were aboard an ancient submarine somewhere off the Philippines. Missile tubes had been converted into fermenters. The bridge was crammed with the latest manipulation technology.

"It's not facilities I need," the pirate told Evan, "it's distribution."

"No problem," said Evan. The pirate's security had been easy to penetrate. He'd tried to infect Evan with a zombie virus, but Evan's designer-immune system had easily dealt with it. Slow poison was so much more subtle: by the time it could be detected it was too late. Evan was 32. He was posing as a Swiss grey market-broker.

"This is where I keep my old stuff," the pirate said, rapping a stainless steel vat. "From before I went big time. Remember when the Brazilian rainforest started to glow? That was me." He dashed sweat from his forehead, frowned at the room's complicated thermostat. Grossly fat and completely hairless, he was wearing only bermuda shorts and shower sandals. He had been targeted because he was about to break the big time with a novel HIV cure. The company was still making big money from its own cure: they ensured AIDS had never been completely eradicated in Third World countries.

The pirate rapped the thermostat with shaking hands. "Hey, is it hot in here, or what?"

"That's the first symptom," Evan said. He stepped sideways as the gene pirate crashed to the deck. "And that's the second."

The company had bought the pirate's security chief. Evan had plenty of time to fix the fermenters. By the time he was ashore, they would have boiled dry. On impulse, against orders, he took a sample of the HIV cure with him.

"The territory between piracy and legitimacy is a minefield," the assassin told Evan. "Definitions shift according to convenience—and that's where I come in. My company likes stability. Another year and your company would have gone public: probably the shares would have made you a billionaire. No-one else has cats these days. We thought their genetic material was eradicated back in the twenties. Very astute of you, going for the luxury market." She frowned. "Why am I talking so much?"

"For the same reason you're not going to kill me," Evan said. "It seems a silly thing to want to do," the assassin admitted. Evan smiled, "I need someone like you in my organization. And since you spent so long getting close enough to seduce me, perhaps you'd do me the honour of becoming my wife. I'll need one."

"You don't mind being married to a killer?"

"Oh, that. I used to be one myself."

Evan saw the market-crash coming. Gene-wars had reduced cereals and many other cash crops to gene sequences stored in computer vaults. Three global companies held patents on the calorific intake of 98 per cent of humanity. But they had lost control of biotechnology. Pressures of the war economy had simplified it to the point where anyone could directly manipulate their own genetic material, and hence alter their own bodies.

Evan had made a fortune in the fashion industry, selling templates and microscopic self-replicating robots which edited DNA. He guessed that someone would eventually devise an artificial photosynthesis system to imitate the way that plants turn organic compounds into food using chlorophyll (which makes plants green) by absorbing sunlight. Evan's stock-market systems were programmed to correlate research in the field. He and his wife sold controlling interests in their company three months before the first green people appeared.

"I remember when you knew what a human being was," Evan said sadly. From her cradle, inside a mist of spray, his wife said, "Is that why you never went green? I always thought it was a fashion statement."

"Old habits die hard." The truth was he liked his body the way it was. These days, going green involved getting body cells to mutate into a metre-high black cowl to absorb sufficient light. Most people lived in the tropics, swarms of black-caped anarchists. Work was no longer a necessity but an indulgence. Evan added, "I'm going to miss you."

"Let's face it," his wife said, "we never were in love. But I'll miss you, too." With a flick of her powerful tail she launched her streamlined body into the sea.

"They wish to honour you by taking your genetic material to Mars," the little purple microsaur said. Evan sighed. "I just want peace. To rest. To die."

"Oh Evan," the microsaur said patiently, "surely even you know that nothing really dies any more."

Noah's Fridge: Putting Earth's Treasures on Ice

AN

EDITORIAL FROM

THE ECONOMIST

The best problems are obvious and pressing. A problem that is daunting but gradual, and complex to boot, is likely to induce worry and helplessness, but little action.

So it could be with Earth's crisis of biodiversity. It is a vast concern. Species are being made extinct at a phenomenal rate. Half of the world's species may be at risk as humans destroy some habitats and pollute others.

But this crisis is also gradual. The world is losing perhaps 1.8 per cent of its forests each year, which is little more than the rate at which a person loses his or her allotted three score years and ten of life.

That rate, quick as lightning in terms of the history of life, slow as old age to an individual human, is one of the reasons why efforts to save the world's species have been so sporadic and feeble. Another is that the various causes—mostly a form of genocidal thoughtlessness as natural habitats and ecosystems are whittled away—often stem from implacable economic and demographic forces: rising populations and the search for higher living standards.

If solving the biodiversity crisis means first dealing with those underlying problems, then solutions will be slow. And all delay means loss. That is why a quick and simple, if only partial, solution would be attractive.

Believe it or not, one is available.

Imagine an international fleet of airships cruising the tropics. At a pre-arranged place in the forest one will drop anchor. The crew covers a tree with a net and sprays an insecticide over it, then gathers up the multitude of creatures to be found in its boughs.

Others take cuttings from the tree and the host of other plants that live on it. Yet more sample the organisms that live in its roots, and in the soil, and on the forest floor. They do not study their findings, but instead pack them off to an archive where, frozen in liquid nitrogen, they will be stored.

The same is done to the next tree, and the next, until a stand of a few dozen, maybe a hundred, has been sampled. Then the airship moves on to do the same thing elsewhere.

In this way a sample of the world's biodiversity can be quickly assembled. The frozen vaults would contain examples of all the orders of creatures on Earth, and a goodly proportion of the individual species and their genes. Future biologists would be provided with a storehouse of knowledge. Judging from the progress biology has made in the past 50 years, the amount that may be done with such knowledge will be staggering. It might well include using DNA from the vaults to bring back vanished species, or to create new replacements.

Many conservationists recoil in horror from this idea. How could those frozen vaults take the place of the vibrant forests, the sweeping savannahs, the lucid reefs, the humble ooze of the mudflats, all teeming with life as yet unrecognized and unacknowledged?

They could not. But they could provide a reasonably good scientific sample of Earth's genetic riches for the same sort of price that people are already accustomed to paying for big science projects. Freezing a million species would be cheap compared with travelling to Saturn, or even mapping the human genome.

Gregory Benford, the physicist and writer who thought up the idea of a big freeze, sees it as providing a reserve of wonder for future generations even if, as he would wish, most of the species within it will anyway be conserved in other ways.

Genetic richness is not the only thing worth saving. Ecosystems provide valuable services and subtly improve the quality of life. If full account is taken of these services the ecosystems' value becomes much clearer, and their protection will often, though not always, be justified in terms of real benefits. Beyond that, there is a set of esthetic, moral and spiritual values. The problem is that these are subjective. It is quite possible for reasonable people not to share them, though there are hints that more and more people are coming to value life's richness and sanctity in and of itself.

These arguments can be strong—but even strong arguments sometimes lose. Those who live closest to nature often have the strongest spiritual bond to it, and the most to lose by damaging it, but they are also often those who, just by staying alive, do the most harm.

Feeling that their arguments need strengthening, many conservationists enlist the pure scientific case for biodiversity. By addressing part of the scientific question independently, the big freeze would make that harder. But the loss would not be great. After all, if pragmatism and morality fail, can appeals to science really be expected to tip the balance?

And the gains are real. While marginally weakening the arguments for conservation, the big freeze provides insurance against some of the consequences should those arguments fail.

The big freeze is a practical notion, and deserves to be taken as such. But even if never instigated, it remains a useful "thought experiment," allowing people to sort out the different reasons for their love of nature, the better to act on them.

For those who dislike thought experiments, though, here is a fairy tale:

There were once three gardeners who cared for orchards of rich fruit and meadows of sweet grass. One day, outside the gardens, they found a huge crowd of big people with bigger appetites, people with feet that, in the joy of dancing, would trample all the earth, and with eyes that only truly came to life when they gazed into fires.

They wanted to enter the gardens. One gardener told them how much he loved his garden, and let them in. The next armed herself with a sharpened hoe. And the third ran under the trees and through the meadows, gathering seeds and flowers from every part of the garden. He packed them carefully and took them to a safe place, hoping that one day the dancing people would love the garden too.

All the gardeners were good and brave, but which was the best steward?

The
IWM 1000

∿

BY

ALICIA YÁÑEZ

COSSÍO

nce upon a time, all the professors disappeared, swallowed and digested by a new system. All the centers of learning closed because they were outmoded, and their sites were converted into living quarters swarming with wise, well-organized people who were incapable of creating anything new.

Knowledge was an item that could be bought and sold. A device called the IWM 1000 had been invented. It was the ultimate invention: it brought an entire era to an end. The IWM 1000 was a very small machine, the size of an old scholarly briefcase. It was very easy to use—lightweight and affordable to any person interested in knowing anything. The IWM 1000 contained all human knowledge and all the facts of all the libraries of the ancient and modern world.

Nobody had to take the trouble of learning anything because the machine, which could be hand-carried or put on any piece of household furniture, provided any information to anybody. Its mechanism was so perfect, and the data it gave so precise, that nobody had dared to prove it otherwise. Its operation was so simple that children spent time playing with it. It was an extension of the human brain. Many people would not be separated from it even during the most personal, intimate acts. The more they depended on the machine, the wiser they became.

A great majority, knowing that the facts were so ready at hand, had never touched an IWM 1000, not even out of curiosity. They did not know how to read or write. They were ignorant of the most elementary things, and it did not matter to them. They felt happy at having one less worry, and they enjoyed the other technological advances more. With the IWM 1000, you could write any type of literature, compose music, and even paint pictures. Creative works were disappearing because anybody, with time and sufficient patience, could make any work similar to and even superior to one made by artists of the past without having to exert the brain or feel anything strange or abnormal.

Some people spent time getting information from the IWM 1000 just for the pleasure of knowing something. Some did it to get out of some predicament, and others asked it things of no importance whatsoever, simply for the pleasure of having someone say something to them, even though it might just be something from their trivial, boring world.

"What is *etatex?*"

"What does *hybrid* mean?"

"How do you make a chocolate cake?"

"What does Beethoven's *Pastorale* mean?"

"How many inhabitants are there actually in the world?"

"Who was Viriato?"

"What is the distance from the Earth to Jupiter?"

"How can you get rid of freckles?"

"How many asteroids have been discovered this year?"

"What is the function of the pancreas?"

"When was the last world war?"

"How old is my neighbor?"

"What does *reciprocal* mean?"

Modulations of the voice fell on some supersensitive electronic membrane, connected with the brain of the machine, and computed immediately the requested information, which was not always the same because, according to the tone of voice, the machine computed the data concisely or with necessary references.

Sometimes two intellectuals would start to talk, and, when one of them had a difference of opinion, he or she would consult the machine. The person would present the problem from his or her own perspective, and the machines would talk and talk. Objections were made, and many times these did not come from the intellectuals but from the machines, who tried to convince each other. The two who had begun the discussion would listen, and when they tired of listening, they would be thinking which of the two machines was going to get the last word because of the power of the respective generators.

Lovers would make the machines conjugate all the tenses of the verb to love, and they would listen to romantic songs. In offices and administrative buildings tape-recorded orders were given, and the IWM 1000 would complete the details of the work. Many people got in the habit of talking only to their own machines; therefore, nobody contradicted them because they knew how the machine was going to respond, or because they believed that rivalry could not exist between a machine and a human being. A machine could not accuse anyone of ignorance: they could ask anything.

Many fights and domestic arguments were conducted through the IWM 1000. The contestants would ask the machine to say to their opponent the dirtiest words and the vilest insults at the highest volume. And, when they wanted to make peace, they could make it at once because it was the IWM 1000 and not they who said those words.

People began to feel really bad. They consulted their IWM 1000's, and the machines told them that their organisms could not tolerate one more dose of pep pills because they had reached the limit of their tolerance. In addition, they computed that the possibilities of suicide were on the increase, and that a change in lifestyle had become necessary.

The people wanted to return to the past, but it was too late. Some tried to put aside their IWM 1000, but they felt defenseless. Then they consulted the machines to see if there was some place in the world where there was nothing like the IWM 1000; and the machines gave information and details about a remote place called Takandia. Some people began to dream about Takandia. They gave the IWM 1000 to those who had only an IWM 100. They began to go through a series of strange actions. They went to museums; they spent time in the sections which contained books looking at something that intrigued them a great deal—something that they wanted to have in their hands—little, shabby syllabaries in which the children of past civilizations learned slowly to read poring over symbols, for which they used to attend a designated site called a school. The symbols were called letters; the letters were divided into syllables; and the syllables were made up of vowels and consonants. When the syllables were joined together, they made words, and the words were oral and written... When these ideas became common knowledge, some people were very content again because these were the first facts acquired for themselves and not through the IWM 1000.

Many left the museums to go out to the few antique shops that remained, and they did not stop until they found syllabaries, which went from hand to hand in spite of their high prices. When the people had the syllabaries, they started to decipher them: *a-e-i-o-u, ma me mi mo mu, pa pe pi po pu*. It turned out to be easy and fun. When they knew how to read, they obtained all the books they could. They were few, but they were books: *The Effect of Chlorophyll on Plants, Les Miserables* by Victor Hugo, *One Hundred Recipes from the Kitchen, The History of the Crusades...* They began to read, and, when they could obtain facts for themselves, they began to feel better. They stopped taking pep pills. They tried to communicate their new sensations to their peers. Some looked at them with suspicion and distrust and labelled them lunatics. Then these few people hastened to buy tickets to Takandia.

After a jet, they took a slow boat, then a canoe. They walked many kilometers and arrived at Takandia. There they found themselves surrounded by horrible beings, who did not even wear modest loinclothes. They lived in the tops of trees; they ate raw meat because they were not familiar with fire; and they painted their bodies with vegetable dyes.

The people who had arrived in Takandia realized that, for the first time in their lives, they were among true human beings, and they began to feel happy. They looked for friends; they yelled as the others did; and they began to strip off their clothes and throw them away among the bushes. The natives of Takandia forgot about the visitors for a few minutes to fight over the discarded clothing...

Science Friction With a Machine

BY

MARCEL STRIGBERGER

When I arrived at the vehicle licence office to renew my car plate sticker I noticed something completely different: the option of dealing with a machine instead of lining up and getting served by a civil servant.

The machine, resembling a banking machine, looked appealing. It accepted Visa and had no line. I took the bait.

The machine's screen even flashed bright and colorful messages. The first was, "Welcome. Do you want to use me to: (A) renew your sticker; (B) renew your licence; (C) pay a fine."

I pushed (A) right on the screen.

The machine replied, "Super. Thanks for using me. Look at all those silly people waiting at the counter. Please punch in your plate number."

I cackled as I watched all those other people grumbling in line.

Soon a cartoon of an open filing cabinet came on with a message, "Gone to look into your file. Be back soon."

How cute, I thought. A moment later my personal information flashed across the screen. Unfortunately everyone around me could see it. It had my name, plate number and address, plus some other vital information including my phone number, occupation and my income for the last three years.

The message was, "If all information is correct, press 'Awesome.'" I quickly pressed "Awesome" before my spectators could complete taking their notes. The machine then asked for my credit card. A cartoon lion came on asking me to put the card into his mouth located at the top. I complied. You don't argue with lions.

I then figured that my new sticker would pop out. Right.

The machine first decided to conduct a survey. It asked, "Do you like using this machine" (A) Very much; (B) Sort of; (C) Not really; (D) What I'd like mostly to have in my hands now is a mallet."

I was about to answer (D). I remembered, however, that the lion still had my Visa in his mouth. I didn't want to antagonize the machine and then hear some crunching sounds and a burp.

I pressed (A) of course.

That was a mistake. The machine then continued the survey. "So you like using me? Wow! Would you recommend me to your friends?"

To be safe, I decided to play along with Frankenstein. I pressed choice, "I would love nothing more."

The next question was "Okay, then give me the names and phone numbers of all your friends."

At this point I was wondering if I could get that "mallet" option back again. Fifteen minutes later, Frank said, "Your package is coming."

On screen a pony express rider came galloping through some western town. Suddenly he dismounted and went into the sheriff's office.

The sheriff came out and said "No sticker for you. You have an unpaid parking fine."

I knew that was impossible. As I shouted at the machine a nearby kibitzer pointed to a button on the sheriff's badge which read, "That's impossible."

I gingerly pushed my finger on the sheriff's chest. At this point I almost expected to get charged with police assault. The sheriff answered back, "That's what they all say."

I continued to protest loudly, "When? Where? In Dodge City?"

The sheriff flashed a message saying, "March 11, 1993 in Sudbury." This was followed by a cartoon of a nickel mill with smoke belching out of four tall smokestacks.

I was livid since I was never anywhere near Sudbury in my life. I asked the people nearby if any of them just happened to have a mallet on them.

Suddenly I heard that galloping again. The sheriff appeared and said, "Sorry. It wasn't your parking ticket. Have a nice day."

Finally, the thing spewed out my new sticker and Visa.

As I was leaving, grumbling, a graphic of Mother Teresa came on with the caption reading, "What's he so upset about? Anyone can make a mistake. We're only human."

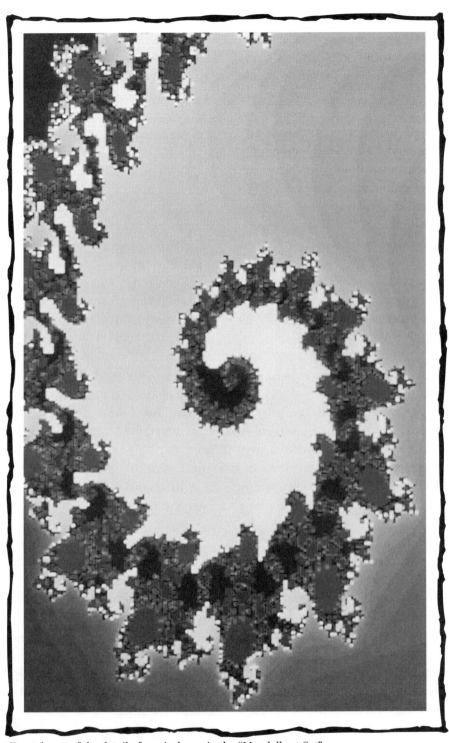

Fractal map of the detail of a spiral arm in the "Mandelbrot Set"

The Sound of One Butterfly

BY

LYNDA HURST

It was only a matter of time before the references started to pop up. And, brace yourselves, here they come.

The Jeff Goldblum character in *Jurassic Park* predicting that the misbegotten dinosaur theme park was an accident waiting to happen?

Why? Chaos theory.

But of course. Pass the popcorn.

Or that chap in Montreal, self-described computer nerd Daniel Corriveau, explaining how he beat staggering odds to win three games in a row (and $600 000) playing electronic keno at the city's casino.

As Corriveau claimed in umpteen celebratory interviews, he had simply monitored the winning sets of numbers over four months and found a pattern in their supposed randomness. From that, he'd worked out the upcoming sequences.

Chaos theory, he said.

Right, well…good on him for beating the system, eh?

Both these characters, the fictional and the real, attempted to explain what on earth they were talking about by invoking an insight that's bucking to become the mantra of the '90s: If a butterfly flutters its wings in Beijing today, it will set off a storm in New York next month.

It's known as the Butterfly Effect, and for the vast majority of the human race, it may be the sole pathway into understanding chaos theory. Or the beginnings of it, at least.

While you may have lived your life happily enough without paying a blind bit of notice to the theory of relativity or, heaven help us, quantum mechanics, you're unlikely to escape chaos. Not only are popular references going to multiply, its disciples say that within a decade, the theory will be taught in the high school curriculum. Right now, it may sound like some end-of-millennium brain game, but over the next 50 years, its implications and applications will dominate mainstream science. Or so it's predicted (though this crowd really should know better).

Back, however, to the beginning.

What that butterfly means to science is that small unknown effects at the start of a process can lead to huge and unforeseeable changes down the road. In weather, for instance, a seemingly negligible disturbance somewhere on the planet will, sooner or later, affect subsequent conditions elsewhere.

Sounds self-evident, to be sure. But there's a little more to it. The changes that are set off will appear random but they are only masquerading as such: chaos does not equal randomness. When they are analyzed via a new computer-imaged form of mathematics called fractal geometry—think shapes, not lines—an underlying structure or pattern can be seen.

In one system, that may mean the potential for a sort of predictability—*x* will happen, we just don't know when. In another, weather the prime example, it may mean knowing for certain there is no long-term predictability. You can only predict that you *can't* predict because...you can never know about the butterfly.

But oh, how humankind yearns to predict, yearns for the order that prediction implies. It's what post-Newtonian science has been all about: finding linear solutions, *rules*, that unlock the mystery of the universe.

Things that didn't fit in were usually ignored as aberrations. Which was easy enough to do in the laboratory but never quite made sense if and when applied to the real world, where straight-line linearity is the exception and chaos the norm.

A few open-minded souls, in particular mathematician Henri Poincair, noticed this wrinkle in the official scheme of things almost a century ago, but didn't pursue the observation. Enter research meteorologist Edward Lorenz.

In 1960, he created a "toy" weather model on a slow and bulky early computer at the Massachusetts Institute of Technology. In it, he fed in daily weather conditions to see what upcoming changes the computer would forecast. Lorenz was aware that classical science said if you knew precisely *all* the starting conditions of a system, you could—in theory—predict its behavior.

So day after day, he'd input the number to signal initial conditions and wait for the laborious printout. One morning, he broke into the computer run to retrieve something, then began the program again. This time, however, he took a shortcut, inputting only a rounded-off number, three decimal places (.506), instead of the earlier six (.506127).

The change was mathematically insignificant, but the effect, to Lorenz's amazement, was enormous. Once printed out, the two programs were seen to deviate dramatically. All because of a tiny—one part in 1000—change in the start-up.

It wasn't news to Lorenz, or indeed any other meteorologist, that there were too many variables for long-term weather predictions ever to be made—though huge amounts of time and money had been invested in the hope.

But the precise *extent* to which it was true—three decimal places in a computer model—a small localized rain shower—was a seismic jolt.

"It may seem obvious on the face of it," says Raymond Kapral, University of Toronto chemistry professor, "that a small change can lead to unpredictable consequences, but it was still a real surprise (to science). The butterfly did make a difference. It was a paradigm shift."

By 1970, the shift had been given the name chaos theory by mathematician James Yorke. The purists didn't like the word because it didn't sound scientific (purists, in fact, didn't like the rule-breaking concept, period).

But the alternative name, non-linear theory, didn't quite do the job. As another mathematician put it, to call the study of chaos "non-linear science" was like calling zoology "the study of non-elephant animals."

In the years since, chaos theory, aided immeasurably by increasingly sophisticated computers, has crept steadily in from the scientific fringes. Along the way, it's upset a few theoretical applecarts, not the least of which has been the most revered of all: Einstein's view of the universe.

Nobody is questioning his science, just his best-known conclusion—that God does not play dice with the universe. Chaoticians insist that "God" does.

What's more, according to American physicist Joseph Ford, "the dice are loaded. And the main objective of physics now is to find out by what rules they were loaded and how we can use them to our own ends."

The new language of chaos is daunting to laypeople. There are the exquisitely beautiful, and quite literally infinite, fractal images that are its hallmark, but there is also period doubling, bifurcations, strange attractors, folded-towel diffeomorphisms, even something called smooth noodle maps. The jargon is daunting even to scientists who didn't cut their teeth on higher mathematics, or who parted company with physics about the same time as most of the rest of the race, somewhere in Grade 10.

But chaos theory has slowly gained converts in all disciplines: mathematics and physics, to be sure, but medicine, biology, ecology, epidemiology and engineering, even economics and politics.

"What's exciting," says U of T physicist Rashni Desai, "is that it gives us a common mathematical language to use in all our diverse disciplines. Before, we were becoming over-specialized, everyone was rediscovering the wheel."

Or in some cases, running up against a wall. As the revolutionary cosmologist and physicist Stephen Hawking said in a 1980 lecture, titled *Is the End in Sight for Theoretical Physics?*, "We already know the physical laws that govern everything we experience in everyday life..."

But even they hadn't explained the central mystery of the whole, not just its parts; the ultimate truth, the Grand Unified Theory...the theory of everything. Chaos probably won't either, not on its own. But it has opened up, globalized and breathed new life into the ongoing attempt.

Wide-ranging specialist journals have been full of its implications for the past few years. There's even been an American bestseller, *Chaos: Making a New Science,* in which author James Gleick can barely contain his enthusiasm for a breakthrough he ranks in importance right up there with relativity and quantum theory.

"Now that science is looking," he writes, "chaos seems to be everywhere. A rising column of cigarette smoke breaks into wild swirls. A flag snaps back and forth. A dripping faucet goes from a steady pattern to a random one. Chaos appears in the behavior of the weather...of an airplane in flight...of cars clustering on an expressway...oil flowing in underground pipes."

He says it's already affecting how executives manage workplaces (hierarchies are less pyramid-shaped); how astronomers look at the solar system (it's no longer seen as the perfect example of clockwork order); how political theorists talk about the stresses leading to conflict (they now know a seemingly insignificant societal change can have unforeseeable, perhaps catastrophic consequences).

Gleick says chaos may be the key to understanding everything from the spreading of rumors to the otherwise inexplicable rise and fall of certain animal species to the cycling of the stock market....

Cornell University mathematician John Guckenheimer urges a soupçon of caution. Yes, he says, chaos theory is important because it provides a means of seeing patterns within disorder that don't show up in traditional tests; means which can be applied to some "systems" in the real world. Maybe, who knows?, the stock market will be one of them.

"But people like Gleick have oversold it," he adds emphatically. "Chaos is no panacea. It doesn't explain what we don't understand. It is just an oversight."

Chaos evangelists counter that, in time, it will prove to be more than an insight. Cardiologist Vivan Rambihar, president of the fledgling Toronto Chaos Society, says the theory will be the scientific approach of the first half of the 21st century because it is a better model with which to understand nature—"It more accurately reflects reality."

Already, real-life clinical applications are under study, he says, with the most promising research being done on the inherent chaos in heart failure, specifically arrhythmia—irregular heartbeats that in their most severe form, ventricle fibrillation, can be suddenly fatal.

Traditional electrocardiograms give only a one-dimensional record where doctors identify the pattern and thus the category of arrhythmia. But the classifications are often superficial. The tools of chaos may be able to help them analyze in detail the dynamics of the pattern.

Pacemakers are already being improved thanks to chaos, says Rambihar, and eventually there may be a means of identifying, sufficiently in advance, those individuals—several hundred thousand a year in North America—at risk of fibrillation.

"Technical applications such as this will give the theory credibility," he says with assurance. Credibility to still-leery scientists and bemused laypersons alike. Even an energetic convert like Rambihar admits that on first acquaintance, chaos theory resembles nothing so much as plain old common sense. But just stay with it. Like his young daughter has. Her entry in a recent school science fair was called *Jurassic Camel: The Last Straw* *(That Broke the Camel's Back)*. The last straw is the butterfly's wings.

A PRIMER ON CHAOS THEORY

A verse by Benjamin Franklin illustrates the importance of seeming trifles:

For want of a nail the shoe was lost,
For want of a shoe the horse was lost,
For want of a horse the rider was lost,
For want of a rider the battle was lost,
For want of a battle the kingdom was lost,
And all for want of a horse shoe nail.

Fear

BY

GREG

NELSON

CHARACTERS   Smith and Jones are of either sex.

SETTING   A public place.

(Smith and Jones are alone. Jones is studying several sheets of paper.)
Smith: Excuse me.
Jones: Yes?
Smith: I'm frightened.
Jones: Sorry?
Smith: I'm frightened.

(Slight pause.)

Jones: Uh…
Smith: I'm supposed to tell someone. If I acknowledge it and name it and put it into words then I can begin to heal. My wounds of fear can begin to heal. Do you mind?
Jones: Of course not.
Smith: Thank you. I don't want to impose.
Jones: Not at all.
Smith: Thank you.

(Pause)

Jones: What are you frightened of?
Smith: I'm not exactly sure.
Jones: Oh.
Smith: Which is frightening, don't you think?

Jones: Yes.

Smith: You do?

Jones: Yes, I do.

Smith: I haven't been sleeping much either. Every time I close my eyes the world starts spinning. Which is ironic I mean because that's what it's doing right? Spinning.

Jones: Yes.

Smith: I mean it's actually spinning. The entire world is spinning all the time.

Jones: Right.

Smith: That's what it's doing.

Jones: On its axis.

Smith: That's right, on its axis, that's right. Spinning.

Jones: Yes.

(Longer pause. Jones goes back to reading.)

Smith: People don't think anymore.

Jones: Sorry?

Smith: They don't. They just sit around and, and watch TV. Why don't people think?

Jones: Well—

Smith: When was the last time you saw somebody really thinking about something, I mean *really* really thinking, I don't mean making a choice, like at the store, I mean actually using their *mind*, actually thinking about some kind of major philosophical or ethical *dilemma?*

Jones: I can't remember.

Smith: You don't see that anymore. I mean people didn't used to go to movies. They used to go to, you know, four-hour-long political debates, you know, sitting there eating their popcorn. Hundred years ago, your average sentence was like a *paragraph* long, today what are they?

Jones: Well—

Smith: Couple of words. Headlines. People don't read anymore.

Jones: That's true.

Smith: They, they, they, they glance.

Jones: Yes.

Smith: People, when they have voices in their heads, they're not *their* voices, they're like Bill Cosby, or, or Oprah, or that guy on the CBC. I mean when they think. Their thoughts. People aren't thinking their own thoughts, that's what I'm saying, they're letting Oprah do their thinking for them, and not just Oprah, Oprah's voice, they're thinking with Oprah's voice, which isn't even thinking, I mean those aren't thoughts, they're just voices they're like, like, meaningless voices, I mean that's not thinking. Why isn't anyone thinking?

Jones: I don't know.

Smith: I personally only watch the news channel. And only then when I can't sleep. Once I get started watching it usually I can't stop, because I get sort of hypnotized by all the pretty colours. And then suddenly it's morning and I still haven't slept and the world's still spinning.

Jones: Look—

Smith: I've been doing some research and frankly I don't give this world ten years. I give it five. Five years. Because we're locked inside this self-destructing system and its going to eat us all alive. Because we can't stop it, we've thrown away the key, we have to wait for it to run it's course, until it drags us down, our entire civilization, until it destroys us, consumes us, sucks us into the black hole of profit and Greed-with-a-capital-"G" until some day soon somebody's going to light a match and the whole thing goes up. The whole thing. All of it. It goes up.

(Jones gives a nervous laugh.)

It's not funny.

Jones: No, I—

Smith: Nobody laughs when they're dead.

(Pause.)

Jones: Do you work for the government?

Smith: What?

Jones: You look familiar.

Smith: What?

Jones: I work for the government. I thought perhaps I might have seen you somewhere. In the corridors.

(Pause. Smith looks at Jones.)

Smith: You work for the government?

Jones: That's right. I type things into computers.

Smith: Oh.

Jones: Secret things.

Smith: Oh.

Jones: There are specific rules. I must not read what I type. I must not understand what I type. I must not remember what I type. After I am finished I must shred what I type.

Smith: Shred...

Jones: Yes, in the shredder. But I don't. For several years now I have in fact refrained from shredding. I have kept these secret documents. I have taken them home and I have read them.

Smith: Oh.

Jones: And I have developed a theory.

(Jones looks around, cautiously, then speaks confidentially to Smith.)

I believe these documents are issued at the very top. I believe these papers are the *source.*

Smith: The source?

Jones: The *source.*

Smith: Of what?

Jones: Of *everything.*

Smith: I see.

Jones: And this is the result. *(Gestures around.)* All of this. This world. Trees, cars, buildings, people, thoughts, *deeds,* all have their source in these papers. Nothing happens by chance. Everything has been antici-pated, indeed, *directed* by these papers.

Smith: Everything?

Jones: *Everything.* I believe these papers are the *key.* I believe that if you could read these papers, you could understand for example your *fears,* and, having understood them, you could calm them.

Smith: If I could ...

Jones: These papers, yes.

(They look at the papers.)

Would you like to have a glance? Would you like to read them?

Smith: Yes.

(Jones gives Smith the papers. Smith tries to read them, fails. Tries again, fails gain.)

Jones: Well?

Smith: I can't read them.

Jones: What?

Smith: They don't make any sense.

Jones: You can't read them?

Smith: No.

Jones: Not at all?

Smith: *(Frightened.)* They don't make any sense.

(Jones takes the papers back.)

What do they say?

(No answer.)

Can you understand them?

(No answer.)

Oh my God. You can't understand them either?

(Jones shakes head.)

Oh my God. That's frightening. I mean I find that really terrifying. Really. Isn't that terrifying?
Jones: Yes. Yes it is.

(Long pause.)

Smith: Sometimes I get the feeling that everything I say has been said before. By someone else. Do you ever get that feeling?
Jones: Yes.

(The end.)

The
Possession

~

BY

SINDIWE

MAGONA

Of course, the incident would not appear in the report on the Conference. How could it? Not one of the delegates understood what exactly had happened. But each would remember the amazing event with much misgiving. And some would send deputies next time there was a conference instead of attending themselves. And the next time. And the next.

It was a Conference to Right the Wrongs of a World Woefully Out of Joint. They came from all corners of the globe to do their duty. Came in their slick, dark suits and their chauffeured limousines; sat with their delegations and acolytes in the Marble Hall; nodded in corridors to faces familiar from the economic shindig in Bangkok or the environmental hoopla in Addis. They gathered to represent the views of their own peoples, to consider the weighty questions before them with the requisite depth of furrow in their brows.

But then something else happened, which pulled their glazed eyes towards the pudgy representative of Outer Manedristan. Just as he was entering the third sub-section, first paragraph of his peroration, he seemed to be taken over by another voice.

"Tell us we will be widows. Tell us we will raise our children all by ourselves. Tell us this when we are little girls. And then help us grow up strong and able to be widows who bring up children all alone. Arm us for this eventuality. Otherwise, what is the use of all this knowledge you gather, if it does not prepare the girl-child for her fate?"

The voice was that of a very old woman. No doubt about that; strong though it was. Firm, loud, unwavering. But it was still the voice of an old woman; unmistakable. Chairs swivelled as delegates craned their necks this way and that, attempting to align sight and sound. But the speaker who had the floor seemed totally unaware of the stir he was causing. Unruffled, he continued with what he thought was his speech.

"I was a girl. Big brother was sent to school. But I was a girl. Small brother was sent to school. But I was made a wife. I cannot complain. My husband was a good man. When the children came he worked and gave us money for food. But then he died. Now I was a widow. Nobody had told me what to do when I became a widow. The bank asked how could they give credit to a woman who could not write her name? And I had no answer. I beg you, tell all the little girls that they will be widows. That men die earlier than women. And send them, all of them, to school. Tell them to learn to read and write so that when they become widows the banks will trust them and give them loans. Little girls, write...your...names. Learn! Learn to write your names."

Without the slightest pause, the speaker continued with what he had been saying before the interruption by the old woman's voice. His droning tones resumed their diplomatic, circuitous tour around his subject: *Abuse of human rights in some Third World countries.*

Eyes which had met uneasily during the strange interruption now reverted to their normal glazed tour of the upper reaches of the marble pillars, the cherubs daubed on the ceiling. We must have imagined it. Or perhaps the learned speaker from Outer Manedristan had simply had a brainstorm, poor chap. At any rate normal service has been resumed now.

But only until the second hijacking of a speaker. (That is what those who dared give a name to what happened at the Conference called it: hijacking). The delegate from West Axum, a smooth-tongued former president with a smug air, was just getting into his stride when a different voice entirely started to emerge from his lips. This time, the voice had a lisp. It was hesitant and shy. A little girl's voice. So frail. She did not sound afraid. No. Just sad.

"Study war no more. Shut your big houses where death is manufactured. Seal the windows. Lock the doors of these terrible houses.

Give the rich men who own them jobs in hospitals and bush clinics where they can see the results of their killing machines. Send them to work in the world's graveyards where they can read the sad tomb-stones of men, women and children killed in their games. Perhaps then other children will not lose their parents as I have, will not lose their homes. Perhaps they will even have money to go to school and food to eat. And their days will not be swathed in fear."

If it had not been so disturbing it would have been a source of some merriment to hear such sentiments expressed by an elder states-man whose own military budget had once been the envy and worry of his neighbours. As it was, a sort of blank hysteria descended upon the delegates: no-one dared to break the spell, to be the one to stand up and protest at this bizarre gate-crashing of the meditations of the mighty.

Simultaneous translation suffered a real blow. Each "intruder" was understood by all there present without recourse to their earpieces. It was a kind of Babel. Only, much, much more. Babel never boasted so many languages as this.

The baby said not a word. How could he? So weak, his faint mewl-ing took the little energy he had. Malnutrition does that, you know. But the mewling was enough. That sound. It went to the hearts of all assembled in the Great Hall of Delegates. And found in each some ancient vestige of human kindness. Granted, a morsel...but still. And the message was more poignant coming through the rather fleshy lips of the obese delegate from one of the richest countries in the world.

"The secret code is broken," said the next intruder, rather threat-eningly. "They have heard of your meeting. The lives you supposedly represent. You get fat while they die. So they have come to deliver their own messages and remind you that they are there, that they are watching as you disport yourselves."

Greed. War. Fear. Hatred. Hunger. Those are just some of the things the voices talked about. The village woman. The orphaned girl. The unborn, the mother, and the worker. Barefooted children walking hungry to school and the song of the hunchback scouring the country-side looking for work.

"Close down the meeting halls and the hotels and the restaurants," said the last of the invaders. "Use the money you waste on meetings to ease the lives that grow daily desperate. Go back to the community. So that your eyes will be opened. So that your hearts inside your bodies will grow and grow. And listen to the prayers of the people. Listen to what we say we need. And don't come and give us needs we didn't even know were there. Can't you see, if the belly isn't grumbling, it can't be hungry? Please, listen. LISTEN!"

The delegates listened all right. They had never listened to a series of speeches with such concentration before. And for years after, in the darkest part of the night they would wake up with a start and remember how those voices reached down deep into their guts and twisted with their truth. They would look with doubt on the assumption that governed their lives, that they were doing their duty by their nation, that they were passing their lives in service. But the moment would pass. They would stroll across to their windows and see the pleasant gardens on the other side of the glass, the lawns and flowerbeds stretching down to the security fence and the night guard's hut. And they would curse themselves for their idiocy. After all, no-one else had ever mentioned what had happened. Perhaps it had been no more than a bad dream.

They would open the drawer and pull out the much-thumbed official record of the Conference. There were the unsullied pearls of the delegate from Outer Manedristan, down in black and white without any intrusion. They would sigh with relief to see that the Conference had been no different from all the other conferences and summit meetings of their eminent careers. And they would return to their beds, thanking their lucky stars as they sank into peaceful oblivion for the foresight of the organizer who had taken the precaution of closing that particular session to the members of the press.

from

Six Degrees of Separation

~

BY

JOHN

GUARE

. . . .

Through flashbacks, this play reveals the story of Paul, a young con artist in New York. He appears at the apartment of a wealthy couple, Ouisa and Flan Kittredge, claiming that he knows their son at college and that he has just been mugged. After inviting Paul in, Flan, Ouisa and their friend Geoffrey have an evening of fascinating conversation with him. In this excerpt, Paul is describing his college thesis.

Flan: *(To us)* And then we asked him what his thesis was on.
Geoffrey: The one that was stolen. Please?
Paul: Well... A substitute teacher out on Long Island was dropped from his job for fighting with a student. A few weeks later, the teacher returned to the classroom, shot the student unsuccessfully, held the class hostage and then shot himself. Successfully. This fact caught my eye: last sentence. *Times.* A neighbor described him as a nice boy. Always reading *Catcher in the Rye.*

The nitwit—Chapman—who shot John Lennon said he did it because he wanted to draw the attention of the world to *Catcher in the Rye* the reading of that book would be his defense.

And young Hinckley, the whiz kid who shot Reagan and his press secretary, said if you want my defense all you have to do is read *Catcher in the Rye*. It seemed to be time to read it again.

Flan: I haven't read it in years. *(Ouisa shushes Flan.)*

Paul: I borrowed a copy from a young friend of mine because I wanted to see what she had underlined and I read this book to find out why this touching, beautiful, sensitive story published in July 1951 had turned into this manifesto of hate.

I started reading. It's exactly as I remembered. Everybody's a phony. Page two: "My brother's in Hollywood being a prostitute." Page three: "What a phony slob his father was." Page nine: "People never notice anything."

Then on page twenty-two my hair stood up. Remember Holden Caulfield—the definitive sensitive youth—wearing his red hunter's cap. "A deer hunter hat? Like hell it is. I sort of closed one eye like I was taking aim at it. This is a people shooting hat. I shoot people in this hat."

Hmmm, I said. This book is preparing people for bigger moments in their lives than I ever dreamed of. Then on page eighty-nine, "I'd rather push a guy out the window or chop his head off with an ax than sock him in the jaw. I hate fist fights…what scares me most is the other guy's face…"

I finished the book. It's a touching story, comic because the boy wants to do so much and can't do anything. Hates all phoniness and only lies to others. Wants everyone to like him, is only hateful, and is completely self-involved. In other words, a pretty accurate picture of a male adolescent.

And what alarms me about that book—not the book so much as the aura about it—is this: the book is primarily about paralysis. The boy can't function. And at the end before he can run away and start a new life, it starts to rain and he folds.

Now there's nothing wrong in writing about emotional and intellectual paralysis. It may indeed, thanks to Chekhov and Samuel Beckett, be the great modern theme.

The extraordinary last lines of *Waiting for Godot*—"Let's go." "Yes, let's go." Stage directions: They do not move.

But the aura around this book of Salinger's—which perhaps should be read by everyone *but* young men—is this: It mirrors like a fun house mirror and amplifies like a distorted speaker one of the great tragedies of our times—the death of the imagination.

Because what else is paralysis?

The imagination has been so debased that imagination—being imaginative—rather than being the linchpin of our existence now stands as a synonym for something outside ourselves like science fiction or some new use for tangerine slices on raw pork chops—what an imaginative summer recipe—and Star Wars! So imaginative and Star Trek—so imaginative! And *Lord of the Rings*—all those dwarfs—so imaginative—

The imagination has moved out of the realm of being our link, our most personal link, with our inner lives and the world outside that world—this world we share. What is schizophrenia but a horrifying state where what's in here doesn't match up with what's out there? Why has imagination become a synonym for style?

I believe that the imagination is the passport we create to take us into the real world.

I believe the imagination is another phrase for what is most uniquely *us*.

Jung says the greatest sin is to be unconscious.

Our boy Holden says "what scares me most is the other guy's face—it wouldn't be so bad if you could both be blindfolded"—most of the time the faces we face are not the other guys' but our own faces. And it's the worst kind of yellowness to be so scared of yourself you put blindfolds on rather than deal with yourself.

To face ourselves.

That's the hard thing.

The imagination.

That's God's gift to make the act of self-examination bearable.

A
New
Departure

~

BY

ALOIS

MOCK

he second World Conference on Human Rights [Vienna, 1993]
brought together all the parties concerned with the
implementation of these rights—governments, United Nations
agencies, specialized and regional international bodies, non-governmental organizations, national institutions, parliamentarians, the
media and private individuals.

The Conference was thus like a living mosaic, reflecting the present
situation of human rights in the world: a lot of pieces need to be polished
and fined down and then put it their proper places to form a meaningful
whole. The Conference was likewise at pains to define, and often to
redefine and reaffirm, principles that have existed for decades or even
centuries. Some of these principles may have given rise to controversy in
the past; the Conference at last gave them universal authority.

Guaranteeing all human beings fundamental rights stemming from
human nature is a long-term undertaking that in the first place entails
reaching agreement among states. What is needed is to establish the
principles of human rights first as political and then as legal obligations,
and finally to see to it that they are put into practice.

The final document of the Conference, consisting of the Declaration and Programme of Action, paves the way for a dynamic development of the UN system to promote and protect human rights in the following fields:

- *The universality of human rights:* the Declaration confirms that all human rights are universal, indivisible and interdependent, even emphasizing that the universal nature of these rights and freedoms is beyond question. It was clear from the preparatory proceedings, and then from the discussions during the Conference, that this principle needed restating.

- *Respect for human rights as a subject for international discussion:* the final document stresses that the promotion and protection of all human rights is a legitimate concern of the international community. Raising questions about the effective respect for these rights can thus no longer be regarded as unwarranted interference in the affairs of a state.

- *The relationship between democracy, development and human rights:* the Conference brought out clearly the interdependence between respect for human rights, economic and social development, and individuals' participation in public life.

- *Minority rights:* the participating states recognized the importance of the promotion and protection of the rights of minorities to the political and social stability of the states in which they live, and accepted the obligation to ensure that persons belonging to minorities may fully and effectively exercise all human rights without discrimination.

- *The position of women and vulnerable groups:* the final document attaches particular importance to the rights of women and of various vulnerable groups, including children, indigenous people, migrant workers and the disabled. It declares that the full and equal participation of women in political, civil, economic, social and cultural life, at the national, regional and international levels, and the eradication of all forms of gender-based discrimination are priority objectives of the international community.

It also reaffirms the commitment of the international community to the economic, social and cultural well-being of indigenous people, and their enjoyment of the fruits of sustainable development.

- *The Programme of Action:* unlike the first World Conference, held at Tehran in 1968, the declaration of the Vienna Conference is combined with a Programme of Action containing guidelines for its implementation in future years.

  The United Nations system is called upon to take concrete measures for the practical application of the recommendations contained in the final document. To this end the Conference asked that the Geneva Centre for Human Rights be strengthened and its financial resources and staff be increased, and that existing machinery for promoting and protecting human rights be made more effective. The proposal that a post of High Commissioner for Human Rights be created has since been adopted by the General Assembly.

  The realization of human rights is, however, an undertaking that is far from complete. The discussions at the Conference confirmed that the modern world is in total disarray. Armed aggression, flagrant and systematic violation of human rights, and the denial of fundamental freedoms continue in various parts of the world. As the Nobel Peace prizewinners gathered at Vienna emphasized during the Conference: "One of the fundamental lessons of our time is that respect for human rights is essential for peace. There can be no real peace without *justice*, and any lasting peace must be founded on universal devotion to the human family. National interests must be subordinated to international obligations....

  "The only way of permanently settling the conflicts still rife throughout the world is by tackling the main causes of human rights violations. Ethnic wars, growing militarism, racial, religious, cultural and ideological hostility, and the denial of social justice will come to an end if all individuals are brought up, educated and trained in a spirit of tolerance based on respect for human rights in accordance with the various instruments relating to human rights adopted by the United Nations system."

The Vienna Declaration and Programme of Action
Adopted 25 June 1993 by the
World Conference on Human Rights

The World Conference on Human Rights,

Recognizing and affirming that all human rights derive from the dignity and worth inherent in the human person, and that the human person is the central subject of human rights and fundamental freedoms, and consequently should be the principal beneficiary and should participate actively in the realization of these rights and freedoms,

Emphasizing the responsibilities of all States, in conformity with the Charter of the United Nations, to develop and encourage respect for human rights and fundamental freedoms for all, without distinction as to race, sex, language or religion,

Considering the major changes taking place on the international scene and the aspirations of all the peoples for an international order based on the principles enshrined in the Charter of the United Nations, including promoting and encouraging respect for human rights and fundamental freedoms for all and respect for the principle of equal rights and self-determination of peoples, peace, democracy, justice, equality, rule of law, pluralism, development, better standards of living and solidarity,

Deeply concerned by various forms of discrimination and violence, to which women continue to be exposed all over the world,

Solemnly adopts the Vienna Declaration and Programme of Action.

1. The World Conference on Human Rights reaffirms the solemn commitment of all States to fulfil their obligations to promote universal respect for, and observation and protection of, all human rights and fundamental freedoms for all in accordance with the Charter of the United Nations, other instruments relating to

human rights, and international law. The universal nature of these rights and freedoms is beyond question.

2. All peoples have the right of self-determination. By virtue of that right they freely determine their political status, and freely pursue their economic, social and cultural development.

    Taking into account the particular situation of peoples under colonial or other forms of alien domination or foreign occupation, the World Conference on Human Rights recognizes the right of peoples to take any legitimate action, in accordance with the Charter of the United Nations, to realize their inalienable right of self-determination. The World Conference on Human Rights considers the denial of the right of self-determination as a violation of human rights and underlines the importance of the effective realization of this right.

5. All human rights are universal, indivisible, and interdependent and interrelated. The international community must treat human rights globally in a fair and equal manner, on the same footing, and with the same emphasis. While the significance of national and regional particularities and various historical, cultural and religious backgrounds must be borne in mind, it is the duty of States, regardless of their political, economic and cultural systems, to promote and protect all human rights and fundamental freedoms.

6. The efforts of the United Nations system towards the universal respect for, and observance of, human rights and fundamental freedoms for all, contribute to the stability and well-being necessary for peaceful and friendly relations among nations, and to improved conditions for peace and security as well as social and economic development, in conformity with the Charter of the United Nations.

8. Democracy, development, and respect for human rights and fundamental freedoms are interdependent and mutually reinforcing. Democracy is based on the freely expressed will of the people to determine their own political, economic, social and cultural systems and their full participation in all aspects of their lives. In the context of the above, the promotion and protection of human

rights and fundamental freedoms at the national and international levels should be universal and conducted without conditions attached. The international community should support the strengthening and promoting of democracy, development, and respect for human rights and fundamental freedoms in the entire world.

9. The World Conference on Human Rights reaffirms that least developed countries committed to the process of democratization and economic reforms, many of which are in Africa, should be supported by the international community in order to succeed in their transition to democracy and economic development.

11. The right to development should be fulfilled so as to meet equitably the developmental and environmental needs of present and future generations. The World Conference on Human Rights recognizes that illicit dumping of toxic and dangerous substances and waste potentially constitutes a serious threat to the human rights to life and health of everyone.

12. The World Conference on Human Rights calls upon the international community to make all efforts to help alleviate the external debt burden of developing countries, in order to supplement the efforts of the Governments of such countries to attain the full realization of the economic, social and cultural rights of their people.

15. Respect for human rights and for fundamental freedoms without distinction of any kind is a fundamental rule of international human rights law. The speedy and comprehensive elimination of all forms of racism and racial discrimination, xenophobia and related intolerance is a priority task for the international community. Governments should take effective measures to prevent and combat them.

17. The acts, methods and practices of terrorism in all its forms and manifestations as well as linkage in some countries to drug trafficking are activities aimed at the destruction of human rights, fundamental freedoms and democracy, threatening territorial integrity, security of State and destabilizing legitimately constituted Governments. The international community should take the

necessary steps to enhance cooperation to prevent and combat terrorism.

18. The human rights of women and of the girl-child are an inalienable, integral and indivisable part of universal human rights. The full and equal participation of women in political, civil, economic, social and cultural life, at the national, regional and international levels, and the eradication of all forms of discrimination on grounds of sex are priority objectives of the international community.

   Gender-based violence and all forms of sexual harassment and exploitation, including those resulting from cultural prejudice and international trafficking, are incompatible with the dignity and worth of the human person, and must be eliminated. This can be achieved by legal measures and through national action and international co-operation in such fields as economic and social development, education, safe maternity and health care, and social support.

   The human rights of women should form an integral part of the United Nations human rights activities, including the promotion of all human rights instruments relating to women.

   The World Conference on Human Rights urges Governments, institutions, intergovernmental and non-governmental organizations to intensify their efforts for the protection and promotion of human rights of women and the girl-child.

19. ...The World Conference on Human Rights reaffirms the obligation of States to ensure that persons belonging to minorities may exercise fully and effectively all human rights and fundamental freedoms without any discrimination and in full equality before the law in accordance with the Declaration on the Rights of Persons Belonging to National or Ethnic, Religious and Linguistic Minorities.

   The persons belonging to minorities have the right to enjoy their own culture, to profess and practise their own religion and to use their own language in private and in public, freely and without interference or any form of discrimination.

20. The World Conference on Human Rights recognizes the inherent dignity and the unique contribution of indigenous people to the development and plurality of society and strongly reaffirms the commitment of the international community to their economic,

social and cultural well-being and their enjoyment of the fruits of sustainable development. States should ensure the full and free participation of indigenous people in all aspects of society, in particular in matters of concern to them.

21. ...National and international mechanisms and programmes should be strengthened for the defence and protection of children, in particular, the girl-child, abandoned children, street children, economically and sexually exploited children, including through child pornography, child prostitution or sale of organs, children victims of diseases including acquired immunodeficiency syndrome, refugee and displaced children, children in detention, children in armed conflict, as well as children victims of famine and drought and other emergencies. International co-operation and solidarity should be promoted to support the implementation of the Convention [on the Rights of the Child] and the rights of the child should be a priority in the United Nations system-wide action on human rights.

22. Special attention needs to be paid to ensuring non-discrimination, and the equal enjoyment of all human rights and fundamental freedoms by disabled persons, including their active participation in all aspects of society.

23. The World Conference on Human Rights reaffirms that everyone, without distinction of any kind, is entitled to the right to seek and to enjoy in other countries asylum from persecution, as well as the right to return to one's own country.

The Old Astronaut

BY

PHILIP

STRATFORD

They put me into orbit
to get a new perspective
on the earth;
at sixteen sunsets
and sunrises a day
I am aging rapidly,
but age is not wisdom.

I can tell you however
that to be free
of earth's atmosphere
is not to be free
of earthly preoccupations;
weightlessness is relative,
and wisdom is not distance.

One part of me would like
to go out on a tangent
into the night forever;
the other pulls me
powerful as gravitation
back to my known home,
but emotional debate is not wisdom.

I register minute changes
in myself: heart-beat, breathing,
blood heat, brain-waves,
flux and reflux, but in itself
the self is not relevant
and all this information
is not wisdom.

The technology that put me here
the rivalry that spurred it
the extravagant energy
and ingenuity expended,
the daring of the enterprise,
all these, at this height,
are not wisdom.

We who have hung
in liberating darkness
above the luminous beauty
of the floating earth,
we know that wisdom is
always to move in concert
with her wandering ways.

The
Astronauts Came From
Corner Brook

~

BY

ALDEN

NOWLAN

e're continually being told that the world has changed more in the past several decades than in all of humanity's previous history. Some commentators on the passing scene have gone so far as to assert that human consciousness is entering a new phase. A few years ago there was What's-his-name's book, *The Greening of America,* which for a month and a half was the single most important volume in the history of western thought. After reading it, a university student told me that Mick Jagger was the Anti-Christ: the thought seemed to please him.

If the traffic isn't too heavy, if the airport isn't blown up and if the plane isn't hijacked, we can breakfast on one side of the Atlantic and dine on the other on the same day. We can switch on the television set and watch a massacre from the comparative safety of our livingrooms; or we can turn on the radio and hear a Canadian singer extol soil conservation, pacifism and free love while imitating Bob Dylan's imitation Oklahoma accent.

Which means that we're all of us inhabitants of one Global Village—or so they say and say and say in every classroom, on every editorial page and on every talk show. Every great thinker from Jane Fonda to Hubert Humphrey, from John Lennon to Merv Griffin, from Johnny Carson to Henry Kissinger, agrees that for homo sapiens it's a whole new ball game.

Today's younger generation is better-informed and more mature than any of its predecessors. Someone—evidently a hermit—came to this conclusion in 1960 and we've been parroting it ever since.

Meanwhile, we go on thinking, feeling and acting in much the same way that human beings have always thought, felt and acted. The most remarkable thing about the Technological Revolution is that there's been no revolution.

Ironically, the theories about the drastic effects of technological change on the human spirit have been developed by the very people least affected by it: artists, teachers and scholars, members of professions which have existed in essentially the same form since the beginning of civilization. The astronauts, on the other hand, were head-scratchin', straw-chewin' good old boys for whom landing on the moon was just another day in the factory. They didn't come from any Global Village; they came, in a manner of speaking, from Corner Brook, Truro, Summerside and Bathurst.

It's been said that a human being is mentally and emotionally capable of relating closely to only a limited number of his or her fellows and that consequently the average inhabitant of a tiny village and the average inhabitant of a great city will know approximately the same number of people.

Similarly, the individual human's capacity for absorbing experience probably has varied little if at all from ancient times to the present. Our ancestors didn't know less than we do, they merely knew different things. If their knowledge was narrower than ours it was also deeper.

What do we learn about the ocean when we cross it in less than the space of a single night? We don't even see it for more than a few minutes and then only through a little window and from a great height. The rest of the time we know it's down there only because we've been told that it's down there. For all the difference it would make to us, we could be flying over an ice-cap, a desert or an enormous sheet of linoleum.

We prattle about the Global Village when many of us don't know the name of the family in the apartment across the hall. The best-informed generation in history includes a great many young men and women who are incapable of forming a coherent sentence in their native language. People refer to themselves as kids even after they've reached an age greater than most of their ancestors' average life expectancy.

This is not to say that we're inferior to those who came before us. That's another fallacy and perhaps a more dangerous one.

But neither are we better. Essentially, we're no different. The world wasn't destroyed and created anew the day the first television station went into operation, although this would appear to be a popular superstition.

The space flights have had immeasurably less effect on humanity's inner world than the invention of the button.

The World Is Too Much with Us

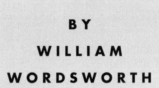

BY

WILLIAM

WORDSWORTH

The world is too much with us; late and soon,
Getting and spending, we lay waste our powers:
Little we see in Nature that is ours;
We have given our hearts away, a sordid boon!
This Sea that bares her bosom to the moon;
The winds that will be howling at all hours,
And are up-gathered now like sleeping flowers;
For this, for everything, we are out of tune;
It moves us not.—Great God! I'd rather be
A Pagan suckled in a creed outworn;
So might I, standing on this pleasant lea,
Have glimpses that would make me less forlorn;
Have sight of Proteus rising from the Sea;
Or hear old Triton blow his wreathèd horn.

Speech to the Young: Speech to the Progress-Toward

BY

GWENDOLYN

BROOKS

Say to them,
say to the down-keepers,
the sun-slappers,
the self-soilers,
the harmony-hushers,
"Even if you are not ready for day
it cannot always be night."
You will be right.
For that is the hard home-run.

Live not for battles won,
Live not for the-end-of-the-song.
Live in the along.

On the Nature of Truth

BY

MAZISI

KUNENE

"People do not follow the same direction, like water"
(Zulu saying)

Those who claim the monopoly of truth
Blinded by their own discoveries of power,
Curb the thrust of their own fierce vision.
For there is not one eye over the universe
But a seething nest of rays ever dividing and ever linking.
The multiple creations do not invite disorder,
Nor are the many languages the enemies of humankind.
But the little tyrant must mould things into one body
To control them and give them his single vision.
Yet those who are truly great
On whom time has bequeathed the gift of wisdom
Know all truth must be born of seeing
And all the various dances of humankind are beautiful
They are enriched by the great songs of our planet.

Sometimes

BY

SHEENAGH

PUGH

Sometimes things don't go, after all,
from bad to worse. Some years, muscadel
faces down frost; green thrives, the crops don't fail,
sometimes a man aims high, and all goes well.

A people sometimes will step back from war;
elect an honest man; decide they care
enough, that they can't leave some stranger poor.
Some men become what they were born for.

Sometimes our best efforts do not go
amiss; sometimes we do as we meant to.
The sun will sometimes melt a field of sorrow
that seemed hard frozen; may it happen for you.

ACKNOWLEDGEMENTS

Care has been taken to trace ownership of copyright material contained in this text. The publishers will gladly accept any information that will enable them to rectify any reference or credit in subsequent editions.

TEXT

p. 1 "Value Earth" by Donella H. Meadows. Reprinted by permission of Donella H. Meadows; p. 4 "In It" by George Johnston, from *The Cruising Auk*. By permission of the author. George Johnston, poet and translator, lives in retirement in Huntingdon, Québec; p. 5 "Together at the Death of a Stranger" by Michele McAlpine. Reprinted by permission of Michele McAlpine; p. 8 "Prayer Before Birth" by Louis MacNeice. From *The Collected Poems of Louis MacNeice* edited by E.R. Dodds and published by Faber and Faber Ltd; p. 10 "The Answer" by Robinson Jeffers. From *The Selected Poetry of Robinson Jeffers* by Robinson Jeffers. Copyright © 1937 and renewed 1965 by Donnan Jeffers and Garth Jeffers. Reprinted by permission of Random House, Inc.; p. 11 "Nausea" by Di Brandt. "Nausea" from *Agnes in the Sky* by Di Brandt, 1990. Used by permission of Turnstone Press, Winnipeg, Canada; p. 17 "The Peace of Wild Things" by Wendell Berry from *Openings*, Copyright 1968 by Wendell Berry, reprinted by permission of Harcourt Brace & Company; p. 19 "Walking" by Linda Hogan. From *Parabola* Magazine, Summer 1988. Copyright © 1988 by Linda Hogan. Reprinted by permission of the author; p. 22 "Exploring the Mysteries of 'Deep Ecology'" by James D. Nations. Reprinted with permission from *Headline News, Science Views*. Copyright © 1991 by the National Academy of Sciences. Courtesy of the National Academy Press, Washington, D.C.; p. 26 "those subjected to radioactivity who did not die were marked for life" by Kristjana Gunnars. Reprinted from *The Night Workers of Ragnarök*, © Kristjana Gunnars 1985, Victoria: Press Porcepic; p. 34 "Lessons From the Third World" by Ben Barber. Reprinted by permission of OMNI, © 1992 OMNI Publications International, Ltd.; p. 36 "It Is Important" by Gail Tremblay is reprinted by permission of the publisher from *Indian Singing in Twentieth Century America* (CALYX Books, © 1990); p. 38 "'Regular Guy' Becomes a Champion" by Frank Jones. Reprinted with permission—The Toronto Star Syndicate; p. 41 "By the Waters of Babylon" by Stephen Vincent Benet. Copyright © 1937 by Stephen Vincent Benet. Copyright © 1965 by Thomas C. Benet, Stephanie B. Mahin, Rachel Benet Lewis. Reprinted by permission of Brandt & Brandt Literary Agents, Inc.; p. 53 "Cosmic Spite" by Grace Nichols. Published by Virago Press Ltd. 1989. Copyright © Grace Nichols 1989. All rights reserved; p. 55 "Sugar Plums and Calabashes" by Bronwen Wallace. From *Arguments with the World: Essays by Bronwen Wallace* edited by Joanne Page, 1992. Reprinted by permission of Quarry Press, Inc.; p. 59 "Mega-City, Mega-Problems" by Andrea Grimaud and p. 66 "Data Bank." Reprinted with permission of *Canada and the World*, Waterloo, Ontario; p. 67 "Top of the Food Chain" copyright © 1993 by T. Coraghessan Boyle, from *Without a Hero* by T. Coraghessan Boyle. Used by permission of Viking Penguin, a division of Penguin Books USA Inc.; p. 74 "The Discovery of Poverty" by Wolfgang Sachs. Reprinted with permission from *New Internationalist* magazine, June 1992 issue; p. 80 "It's Time to Rethink how we Divide The World" by Paul Harrison © Guardian News Service; p. 84 "Why Asia's Tigers Burn So Bright" by John Stackhouse. The Globe and Mail; p. 92 "Malaria Story an Inspiration" by David Suzuki. Reprinted with permission of the author; p. 94

"Children Are Color-Blind" by Genny Lim. "Children are Color-Blind" was first published in *The Forbidden Stitch: An Asian American Women's Anthology*, edited by Shirley Geok-lin Lim, et al., published by CALYX Books © 1989. Reprinted by permission of the publisher; **p. 96** "Revolution!" by Vic Finkelstein. Reprinted with permission from *New Internationalist* magazine, July 1992 issue; **p. 100** "The Cardboard Room" by Teresa Pitman. Reprinted by permission of Teresa Pitman; **p. 107** Excerpt from *Daughters of the Twilight* by Farida Karodia. The extracts from *Daughters of the Twilight* by Farida Karodia, first published by The Women's Press Ltd., 1986, 34 Great Sutton St., London ECIV 0DX, reprinted on pages 107-117 are used by permission of The Women's Press Ltd.; **p. 118** "No Longer Our Own Country" by Tanure Ojaide. © London Magazine; **p. 120** "Walking Both Sides of an Invisible Border" by Alootook Ipellie. Reprinted by permission of Alootook Ipellie; **p. 122** "Slamming Doors" by Ron Redmond. From *Refugees*, No. 88/January 1992, UNHCR; **p. 128** "Elena" by Pat Mora is reprinted with permission from the publishers of *Chants* (Houston: Arte Publico Press—University of Houston, 1985); **p. 129** "The Poverty of Affluence" by Jacinta Goveas. Reprinted with permission from the author. Jacinta Goveas grew up in Pakistan and now works in Toronto's immigrant and refugee community; **p. 135** Untitled poem ("The Earth Is Weary") by Sujeetha Sivarajah © The Toronto Board of Education; **p. 136** "I Was Sixteen" by Rajakumar Thangarajah. From *New Canadian Voices* by Jessie Porter. Reprinted with the permission of the publisher, Wall & Emerson, Inc., Toronto; **p. 138** "Hands of Lead, Feet of Clay" by Jamal Mahjoub. Reprinted with permission from *New Internationalist*, November 1984; **p. 144** "Love and Death in Sarajevo" by Criselda Yabes. Reprinted with permission from *New Internationalist*, May 1993; **p. 149** "Synonyms for War-torn" by Oakland Ross. Reprinted by permission of Cormorant Books, Inc.; **p. 164** "Protection of World's Young" by Amelia A. Newcomb. Reprinted by permission of Amelia A. Newcomb"; **p. 169** "The Children of Bogota" by Patrick Lane. By permission of the author; **p. 170** "Without Hands" by Lorna Crozier. From *Angels of Flesh, Angels of Silence* by Lorna Crozier. Used by permission of the Canadian Publishers, McClelland & Stewart, Toronto; **p. 172** "Our Father Who Art in Heaven" by José Leandro Urbina, translated by Christina Shantz. From *Lost Causes* by José Leandro Urbina, translated by Christina Shantz. Reprinted by permission; **p. 173** "The Purple Children" by Edith Pargeter © Edith Pargeter, 1965, 1994; **p. 180** "Just Lather, That's All" by Hernando Téllez. From *Great Spanish Short Stories* (Dell, 1962). Reprinted by permission; **p. 185** "Third Wave War." Excerpt from *War and Anti-War* by Alvin & Heidi Toffler. Copyright © 1993 by Alvin Toffler and Heidi Toffler. By permission of Little, Brown & Company; **p. 192** "Evan's Progress" by Paul McAuley. Reprinted with permission from *New Internationalist* magazine, March 1991 issue; **p. 197** "Noah's Fridge: Putting Earth's Treasures on Ice" © 1993 The Economist Newspaper Group, Inc. Reprinted with permission; **p. 200** "The IWM 1000" by Alicia Yáñez Cossío is reprinted with permission from the publisher of *Short Stories by Latin American Women: The Magic and the Real*, edited by Celia Correas de Zapata (Houston: Arte Publico Press—University of Houston, 1988); **p. 204** "Science Friction With a Machine" by Marcel Strigberger. By permission of the author. Marcel Strigberger is a Toronto litigation lawyer and freelance humourist; **p. 207** "The Sound of One Butterfly" by Lynda Hurst. Reprinted with permission—The Toronto Star Syndicate; **p. 213** "Fear" by Greg Nelson. "Fear" first appeared in *Instant Applause: 26 Very Complete Plays*, Blizzard Publishing, 1994. Reprinted by permission of the author; **p. 218** "The Possession" by Sindiwe Magona. Reprinted with permission from

New Internationalist magazine, June 1993 issue; **p. 222** From *Six Degrees of Separation* by John Guare. Copyright © 1990 by John Guare. Reprinted by permission of Random House, Inc.; **p. 226** "A New Departure" by Alois Mock. Reprinted from *The UNESCO Courier*, March 1994, Alois Mock; **p. 235** "The Old Astronaut" by Philip Stratford is reprinted from *The Rage of Space* by permission of Oberon Press; **p. 237** "The Astronauts Came From Corner Brook" from *Double Exposure* by Alden Nowlan. "Double Exposure" by Alden Nowlan published by Brunswick Press; **p. 241** "Speech to the Young: Speech to the Progress-Toward" by Gwendolyn Brooks © 1991. Published in *Blacks* by Third World Press, Chicago, 1991; **p. 243** "On the Nature of Truth" by Mazisi Kunene. Reprinted by permission of Mazisi Kunene, recipient of the 1993 Poet Laureate for the Afro Arab Region; **p. 244** "Sometimes" by Sheenagh Pugh from *Selected Poems* (Seren, 1990).

PHOTOGRAPHS

p. viii UNICEF photo by Satyan (UNICEF News, Issue 93); **p. 16** Robert Garrard; **p. 18** Dick Hemingway; **p. 25** Robert Garrard; **p. 54** CIDA/Bruce Paton; **p. 60** CIDA/Pat Morrow; **p. 73** Robert Garrard; **p. 79** United Nations Photo; **p. 91** CIDA/David Barbour; **p. 106** United Nations/Pendl; **p. 133** CALVIN AND HOBBES copyright Watterson. Reprinted with permission of UNIVERSAL PRESS SYNDICATE. All rights reserved; **p. 134** Dick Hemingway; **p. 143** United Nations Photo/J. Isaac; **p. 168** United Nations Photo/M. Graham; **p. 191** Canapress Photo Service; **p. 206** Chris Stone; **p. 234** NASA/WWF; **p. 242** Lynda Powell; **p. 245** Robert Garrard.